Undergraduate Texts in Mathematics

Undergraduate Texts in Mathematics

Apostol: Introduction to Analytic Number Theory.
1976. xii, 338 pages. 24 illus.

Armstrong: Basic Topology.
1983. xii, 260 pages. 132 illus.

Bak/Newman: Complex Analysis.
1982. x, 224 pages. 69 illus.

Banchoff/Wermer: Linear Algebra Through Geometry.
1983. x, 257 pages. 81 illus.

Childs: A Concrete Introduction to Higher Algebra.
1979. xiv, 338 pages. 8 illus.

Chung: Elementary Probability Theory with Stochastic Processes.
1975. xvi, 325 pages. 36 illus.

Croom: Basic Concepts of Algebraic Topology.
1978. x, 177 pages. 46 illus.

Fischer: Intermediate Real Analysis.
1983. xiv, 770 pages. 100 illus.

Fleming: Functions of Several Variables. Second edition.
1977. xi, 411 pages. 96 illus.

Foulds: Optimization Techniques: An Introduction.
1981. xii, 502 pages. 72 illus.

Franklin: Methods of Mathematical Economics. Linear and Nonlinear Programming. Fixed-Point Theorems.
1980. x, 297 pages. 38 illus.

Halmos: Finite-Dimensional Vector Spaces. Second edition.
1974. viii, 200 pages.

Halmos: Naive Set Theory.
1974, vii, 104 pages.

Iooss/Joseph: Elementary Stability and Bifurcation Theory.
1980. xv, 286 pages. 47 illus.

Kemeny/Snell: Finite Markov Chains.
1976. ix, 224 pages. 11 illus.

Lang: Undergraduate Analysis
1983. xiii, 545 pages. 52 illus.

Lax/Burstein/Lax: Calculus with Applications and Computing, Volume 1.
1976. xi, 513 pages. 170 illus.

LeCuyer: College Mathematics with A Programming Language.
1978. xii, 420 pages. 144 illus.

Macki/Strauss: Introduction to Optimal Control Theory.
1981. xiii, 168 pages. 68 illus.

Malitz: Introduction to Mathematical Logic: Set Theory - Computable Functions - Model Theory.
1979. xii, 198 pages. 2 illus.

Martin: The Foundations of Geometry and the Non-Euclidean Plane.
1975. xvi, 509 pages. 263 illus.

Martin: Transformation Geometry: An Introduction to Symmetry.
1982. xii, 237 pages. 209 illus.

Millman/Parker: Geometry: A Metric Approach with Models.
1981. viii, 355 pages. 259 illus.

Owen: A First Course in the Mathematical Foundations of Thermodynamics.
1983, xiv, 178 pages. 52 illus.

continued

John G. Kemeny
J. Laurie Snell

Finite Markov Chains

With a New Appendix
"Generalization of a Fundamental Matrix"

With 12 Illustrations

Springer-Verlag
New York Berlin Heidelberg Tokyo

J. G. Kemeny
Department of Mathematics
Dartmouth College
Hanover, NH 03755
U.S.A.

J. L. Snell
Department of Mathematics
Dartmouth College
Hanover, NH 03755
U.S.A.

AMS Subject Classifications: 60–01, 60J10

Originally published in 1960 by Van Nostrand, Princeton, NJ.

New appendix originally appeared in *Linear Algebra and its Applications*, vol. 38, Elsevier North Holland, Inc., 1981, pp. 193–206. Reprinted with permission.

Library of Congress Cataloging in Publication Data
Kemeny, John G.
Finite Markov Chains.
(Undergraduate texts in mathematics)
Reprint. Originally published: Princeton, NJ:
Van Nostrand, 1960. With new appendix.
1. Markov processes. I. Snell, J. Laurie (James
Laurie), 1925– . II. Title. III. Series.
QA274.7.K45 1983 519.2′33 83–17031

Printed and bound by R. R. Donnelley & Sons, Harrisonburg, VA.
Printed in the United States of America.

9 8 7 6 5 4 3 2

ISBN 0-387-90192-2 Springer-Verlag New York Berlin Heidelberg Tokyo
ISBN 3-540-90192-2 Springer-Verlag Berlin Heidelberg New York Tokyo

PREFACE

The basic concepts of Markov chains were introduced by A. A. Markov in 1907. Since that time Markov chain theory has been developed by a number of leading mathematicians. It is only in very recent times that the importance of Markov chain theory to the social and biological sciences has become recognized. This new interest has, we believe, produced a real need for a treatment, in English, of the basic ideas of finite Markov chains.

By restricting our attention to finite chains, we are able to give quite a complete treatment and in such a way that a minimum amount of mathematical background is needed. For example, we have written the book in such a way that it can be used in an undergraduate probability course, as well as a reference book for workers in fields outside of mathematics.

The restriction of this book to finite chains has made it possible to give simple, closed-form matrix expressions for many quantities usually given as series. It is shown that it suffices for all types of problems to consider just two types of Markov chains, namely absorbing and ergodic chains. A "fundamental matrix" is developed for each type of chain, and the other interesting quantities are obtained from the fundamental matrices by elementary matrix operations.

One of the practical advantages of this new treatment of the subject is that these elementary matrix operations can easily be programed for a high-speed computer. The authors have developed a pair of programs for the IBM 704, one for each type of chain, which will find a number of interesting quantities for a given process directly from the transition matrix. These programs were invaluable in the computation of examples and in the checking of conjectures for theorems.

A significant feature of the new approach is that it makes no use of the theory of eigen-values. The authors found, in each case, that the expressions in matrix form are simpler than the corresponding expressions usually given in terms of eigen-values. This is presumably due to the fact that the fundamental matrices have direct probabilistic interpretations, while the eigen-values do not.

The book falls into three parts. Chapter I is a very brief summary of prerequisites. Chapters II–VI develop the theory of Markov chains. Chapter VII contains applications of this theory to problems in a variety of fields. A summary of the symbols used and of the principal definitions and formulas can be found in the appendices together with page references. Therefore, there is no index, but it is hoped that the detailed table of contents and the appendices will serve a more useful purpose.

It was not intended that Chapter I be read as a unit. The book can be started in Chapter II, and the reader has the option of looking up the brief summary of any prerequisite topic not familiar to him, when he needs it in a later chapter.*

The book was designed so that it can be used as a text for an undergraduate mathematics course. For this reason the proofs were carried out by the most elementary methods possible. The book is suitable for a one-semester course in Markov chains and their applications. Selections from the book (presumably from Chapters II, III, IV, and possibly VII) could also be used as part of an upper-class course in probability theory. For this use, exercises have been given at the end of Chapters II–VI.

The following system of notation has been used in the book: Numbers are denoted by small italic letters, matrices by capital italics, vectors by Greek letters. Functions, sets, and other abstract objects are denoted by boldface letters.

The authors gratefully acknowledge support by the National Science Foundation to the Dartmouth Mathematics Project. Many of the original results in this book were found by the authors while working on this project. The authors are also grateful for computing time made available by the M.I.T. and Dartmouth Computation Center for the development of the above-mentioned programs and for the use of these programs.

The authors wish to express their thanks to two research assistants, P. Perkins and B. Barnes, for many valuable suggestions as well as for their careful reading of the manuscript. Thanks are due to Mrs. M. Andrews and Mrs. H. Hanchett for typing the manuscript.

THE AUTHORS

Hanover, New Hampshire

* A more detailed treatment of most of these topics may be found in one of the following books: (1) *Modern Mathematical Methods and Models*, Volumes 1 and 2, by the Dartmouth Writing Group, published by the Mathematical Association of America, 1958. [Referred to as M^4.] (2) *Introduction to Finite Mathematics*, by Kemeny, Snell, and Thompson, Prentice-Hall, 1957. [Referred to as FM.] (3) *Finite Mathematical Structures*, by Kemeny, Mirkil, Snell, and Thompson, Prentice-Hall, 1959. [Referred to as FMS.] For the prerequisites in probability theory, as well as a treatment of Markov chains from a different point of view, the reader may also wish to consult *Introduction to Probability Theory and Its Applications*, by W. Feller, Wiley, 1957.

PREFACE TO THE SECOND PRINTING

When the authors wrote *Finite Markov Chains*, two fundamental matrices N for absorbing chains and Z for ergodic chains were used to compute the basic descriptive quantities for Markov chains. The choice of N was natural but the choice of Z was less natural. Z was needed to solve equations of the form $(I-P)x = f$ where f is known. Since $I-P$ does not have an inverse, $I-P$ was modified by adding the matrix A, all of whose rows are the fixed probability vector α, and the resulting inverse $Z = (I-P+A)^{-1}$ was used as the fundamental matrix for ergodic chains. This had the disadvantage of having to find α before computing Z.

With the development of pseudo-inverses, it was pointed out by C. D. Meyer* that pseudo-inverses could be used to find basic quantities for ergodic Markov chains, including the fixed vector α. Independently, while teaching Markov chains, Kemeny noticed that it was not necessary to use A and that A could be replaced by any matrix B all of whose rows are the same vector β whose components sum to 1. The resulting matrix $Z = (I-P+B)^{-1}$ serves in many ways the same role as the fundamental matrix used in the book. (Actually, it is sufficient to assume that the sum of the components of β is non-zero.) The fixed vector α is obtained from such a Z by $\alpha = \beta Z$ and certain other basic quantities for ergodic chains, such as the mean first passage times, have the same formula as in the book. In any event, the fundamental matrix used in the book is easily obtained from any matrix in this new class. Kemeny further showed that these results are special cases of a more general theorem in linear algebra.

We have included Kemeny's paper as an appendix to enable the readers to benefit from the simplification resulting from the use of this more general class of fundamental matrices.

* Carl D. Meyer, "The role of the group generalized inverse in the theory of finite Markov chains", *SIAM Rev.* **17**:443–464, 1975.

TABLE OF CONTENTS

CHAPTER I—PREREQUISITES

CHAPTER II—BASIC CONCEPTS OF MARKOV CHAINS

CHAPTER III—ABSORBING MARKOV CHAINS

CHAPTER IV—REGULAR MARKOV CHAINS

CHAPTER V—ERGODIC MARKOV CHAINS

CHAPTER VI—FURTHER RESULTS

CHAPTER VII—APPLICATIONS OF MARKOV CHAINS

CHAPTER I

PREREQUISITES

§ **1.1 Sets.** By a *set* a mathematician means an arbitrary but well-defined collection of objects. Sets will be denoted by bold capital letters. The objects in the collection are called *elements*.

If $\mathbf{A}$ is a set, and $\mathbf{B}$ is a set whose elements are some (but not necessarily all) of the elements of $\mathbf{A}$, then we say that $\mathbf{B}$ is a *subset* of $\mathbf{A}$, symbolized as $\mathbf{B} \subseteq \mathbf{A}$. If the two sets have exactly the same elements, then we say that they are equal, i.e. $\mathbf{A} = \mathbf{B}$. Thus $\mathbf{A} = \mathbf{B}$ if and only if $\mathbf{A} \subseteq \mathbf{B}$ and $\mathbf{B} \subseteq \mathbf{A}$. If $\mathbf{B}$ is a subset of $\mathbf{A}$ and is not equal to $\mathbf{A}$, then we say that it is a *proper subset*, and write $\mathbf{B} \subset \mathbf{A}$. If $\mathbf{A}$ and $\mathbf{B}$ have no element in common, we say that they are *disjoint*.

Very frequently we will deal with a given set of objects, and discuss various subsets of it. The entire set will be called the *universe*, $\mathbf{U}$. A particularly interesting subset is the set with no elements, the *empty set* $\mathbf{E}$.

Given a set, there are a number of ways of getting new subsets from old ones. If $\mathbf{A}$ and $\mathbf{B}$ are both subsets of $\mathbf{U}$, then we define the following operations:

(1) The *complement* of $\mathbf{A}$, $\tilde{\mathbf{A}}$, has as elements all the elements of $\mathbf{U}$ which are not in $\mathbf{A}$.

(2) The *union* of $\mathbf{A}$ and $\mathbf{B}$, $\mathbf{A} \cup \mathbf{B}$, has as elements all the elements of $\mathbf{A}$ and all the elements of $\mathbf{B}$.

(3) The *intersection* of $\mathbf{A}$ and $\mathbf{B}$, $\mathbf{A} \cap \mathbf{B}$, has as elements all the elements that $\mathbf{A}$ and $\mathbf{B}$ have in common.

(4) The difference of $\mathbf{A}$ and $\mathbf{B}$, $\mathbf{A} - \mathbf{B}$ has as elements all the elements of $\mathbf{A}$ that are not in $\mathbf{B}$.

To illustrate these operations, we will list some easily provable relations between these sets:

$$\tilde{\mathbf{U}} = \mathbf{E} \qquad \widetilde{\mathbf{A} \cup \mathbf{B}} = \tilde{\mathbf{A}} \cap \tilde{\mathbf{B}} \qquad \mathbf{A} \cap \mathbf{B} = \mathbf{B} \cap \mathbf{A}$$

$$\tilde{\tilde{\mathbf{A}}} = \mathbf{A} \qquad \widetilde{\mathbf{A} \cap \mathbf{B}} = \tilde{\mathbf{A}} \cup \tilde{\mathbf{B}} \qquad \mathbf{A} \cup \mathbf{E} = \mathbf{A}$$

$$\mathbf{A} - \mathbf{B} = \mathbf{A} \cap \tilde{\mathbf{B}} \qquad \mathbf{A} \cup \mathbf{B} = \mathbf{B} \cup \mathbf{A} \qquad \mathbf{A} \cap \mathbf{E} = \mathbf{E}$$

If $\mathbf{A}_1, \mathbf{A}_2, \ldots, \mathbf{A}_r$ are subsets of **U**, and every element of **U** is in one and only one set $\mathbf{A}_j$, then we say that $\mathbf{A} = \{\mathbf{A}_1, \mathbf{A}_2, \ldots, \mathbf{A}_r\}$ is a *partition* of **U**.

If we wish to specify a set by listing its elements, we write the elements inside curly brackets. Thus, for example, the set of the first five positive integers is $\{1, 2, 3, 4, 5\}$. The set $\{1, 3, 5\}$ is a proper subset of it. The set $\{2\}$, which is also a subset of the five-element set, is called a *unit set*, since it has only one element.

In the course of this book we will have to deal with both finite and infinite sets, i.e. with sets having a finite number or an infinite number of elements. The only infinite sets that are used repeatedly are the set of integers $\{1, 2, 3, \ldots\}$ and certain simple subsets of this set.

For a more detailed account of the theory of sets see FM Chapter II or FMS Chapter II.†

§ **1.2 Statements.** We are concerned with a process which will frequently be a scientific experiment or a game of chance. There are a number of different possible outcomes, and we will consider various statements about the outcome.

We form the set **U** of all logically possible outcomes. These must be so chosen that we are assured that exactly one of these will take place. The set **U** is called the *possibility space*. If **p** is any statement about the outcome, then it will (in general) be true according to some possibilities, and false according to others. The set **P** of all possibilities which would make **p** true is called the *truth set* of **p**. Thus to each statement about the outcome we assign a subset of **U** as a truth set. The choice of **U** for a given experiment is not unique. For example, for two tosses of a coin we may analyze the possibilities as **U** = {**HH**, **HT**, **TH**, **TT**} or **U** = {0**H**, 1**H**, 2**H**}. In the first case we give the outcome of each toss and in the second only the number of heads which turn up. (For a more detailed discussion of this concept see FM Chapter II or FMS Chapter II.)

Given two statements **p** and **q** having the same subject matter (i.e. the same **U**), we have a number of ways of forming new statements from them. (We will assume that the statements have **P** and **Q** as truth sets:)

(1) The statement $\sim\mathbf{p}$ (read "not **p**") is true if and only if **p** is false. Hence it has $\tilde{\mathbf{P}}$ as truth set.

(2) The statement $\mathbf{p} \vee \mathbf{q}$ (read "**p** or **q**") is true if either **p** is true or **q** is true or both. Hence it has $\mathbf{P} \cup \mathbf{Q}$ as truth set.

† FM = Kemeny, Snell, and Thompson, *Introduction to Finite Mathematics*, Englewood Cliffs, N.J., Prentice-Hall, Inc., 1957.

FMS = Kemeny, Mirkil, Snell, and Thompson, *Finite Mathematical Structures*, Englewood Cliffs, N.J., Prentice-Hall, Inc., 1959.

(3) The statement $\mathbf{p} \wedge \mathbf{q}$ (read "**p** and **q**") is true if both **p** and **q** are true. Hence it has $\mathbf{P} \cap \mathbf{Q}$ as truth set.

Two special kinds of statements are among the principal concerns of logic. A statement that is true for each logically possible outcome, that is, a statement having **U** as its truth set, is said to be *logically true* (such a statement is sometimes called a tautology). A statement that is false for each logically possible outcome, that is a statement having **E** as its truth set, is *logically false* or *self-contradictory*.

Two statements are said to be *equivalent* if they have the same truth set. That means that one is true if and only if the other is true.

The statements $\mathbf{p}_1, \mathbf{p}_2, \ldots, \mathbf{p}_k$ are *inconsistent* if the intersection of their truth sets is empty, i.e., $\mathbf{P}_1 \cap \mathbf{P}_2 \cap \cdots \cap \mathbf{P}_k = \mathbf{E}$. Otherwise they are said to be *consistent.* If the statements are inconsistent, then they cannot all be true. If they are consistent, then they could all be true.

The statements $\mathbf{p}_1, \mathbf{p}_2, \ldots, \mathbf{p}_k$ are said to form a *complete set of alternatives* if for every element of **U** exactly one of them is true. This means that the intersection of any two truth sets is empty, and the union of all the truth sets is **U**. Thus the truth sets of a complete set of alternatives form a partition of **U**. A complete set of alternatives provides a new way (and normally a less detailed way) of analyzing the possible outcomes.

§ **1.3 Order relations.** We will need some simple ideas from the theory of order relations. A complete treatment of this theory will be found in M[4], Vol. II, Unit 2.† We will take only a few concepts from that treatment.

Let **R** be a relation between two objects (selected from a specified set **U**). We denote by **aRb** the fact that **a** holds the relation **R** to **b**. Some special properties of such relations are of interest to us.

1.3.1 DEFINITION. *The relation* **R** *is* reflexive *if* **xRx** *holds for all* **x** *in* **U**.

1.3.2 DEFINITION. *The relation* **R** *is* symmetric *if whenever* **xRy** *holds, then* **yRx** *also holds, for all* **x**, **y** *in* **U**.

1.3.3 DEFINITION. *The relation* **R** *is* transitive *if whenever* $\mathbf{xRy} \wedge \mathbf{yRz}$ *holds, then* **xRz** *also holds, for all* **x**, **y**, **z** *in* **U**.

1.3.4 DEFINITION. *A relation that is reflexive, symmetric, and transitive is an* equivalence relation.

The fundamental property of an equivalence relation is that it partitions the set **U**. More specifically, let us suppose that **R** is an

† M[4] = *Modern Mathematical Methods and Models*, by the Dartmouth Writing Group. Mathematical Association of America, 1958.

equivalence relation defined on **U**. We put elements of **U** into classes in such a manner that two elements **a** and **b** are in the same class if **aRb**. It can be shown that the resulting classes are well defined and mutually exclusive, giving us a partition of **U**. These classes are the *equivalence classes* of **R**.

For example, let **xRy** express that "**x** is the same height as **y**," where **U** is a set of human beings. Then the resulting partition divides these people according to their heights. Two men are in the same equivalence class if and only if they are the same height.

1.3.5 DEFINITION. *A relation* **T** *is said to be* consistent *with the equivalence relation* **R** *if, given that* **xRy**, *then if* **xTz** *holds so does* **yTz**, *and if* **zTx** *holds so does* **zTy**.

1.3.6 DEFINITION. *A relation that is reflexive and transitive is known as a* weak ordering relation.

A weak ordering relation can be used to order the elements of **U**. Given a weak ordering **T**, and given any two elements **a** and **b** of **U**, there are four possibilities: (1) $\mathbf{aTb} \wedge \mathbf{bTa}$; then the two elements are "alike" according to **T**. (2) $\mathbf{aTb} \wedge \sim(\mathbf{bTa})$; then **a** is "ahead" of **b**. (3) $\sim(\mathbf{aTb}) \wedge \mathbf{bTa}$; then **b** is "ahead." (4) $\sim(\mathbf{aTb}) \wedge \sim(\mathbf{bTa})$; then we are unable to compare the two objects.

For example, if **xTy** expresses that "I like **x** at least as well as **y**," then the four cases correspond to "I like them equally," "I prefer **x**," "I prefer **y**," and "I cannot choose," respectively.

The relation of being alike acts as an equivalence relation. Indeed, it can be shown that if **T** is a weak ordering, then the relation **xRy** that expresses that $\mathbf{xTy} \wedge \mathbf{yTx}$ is an equivalence relation consistent with **T**. Thus **T** serves both to classify and to order. Consistency assures us that equivalent elements of **U** have the same place in the ordering.

For example, if we choose "is at least as tall" as our weak ordering, this determines the equivalence relation "is the same height," which is consistent with the original relation.

1.3.7 DEFINITION. *If* **T** *is a weak ordering, then the relation* $\mathbf{xTy} \wedge \mathbf{yTx}$ *is the equivalence relation* determined *by it.*

1.3.8 DEFINITION. *If* **T** *is a weak ordering, and the equivalence relation determined by it is the identity relation* ($\mathbf{x}=\mathbf{y}$) *then* **T** *is a* partial ordering.

The significance of a partial ordering is that no two distinct elements are alike according to it. One simple way of getting a partial ordering is as follows: Let **T** be a weak ordering defined on **U**. Define a new relation **T*** on the set of equivalence classes by saying that **uT*v** holds if every element of **u** bears the relation **T** to every element of **v**.

This is a partial ordering of the equivalence classes, and we call it the partial ordering *induced* by **T**.

1.3.9 DEFINITION. *An element* **a** *of* **U** *is called a* minimal element *if* **aTx** *implies* **xTa** *for all* $\mathbf{x} \in \mathbf{U}$. *If a minimal element is unique, we call it a* minimum.

We can define "maximal element" and "maximum" similarly. If U is a finite set, then it is easily shown that for any weak ordering there must be at least one minimal element. However, this minimal element need not be unique. Similarly, the weak ordering must have a maximal element, but not necessarily a maximum.

§ **1.4 Communication relations.** An important application of order relations is the study of communication networks. Let us suppose that **r** individuals are connected through a complex network. Each individual can pass a message on to a subset of the individuals. This we will call *direct contact.* These messages may be relayed, and relayed again, etc. This will be *indirect contact.* It will *not* be assumed that a member can contact himself directly. Let **aTb** express that the individual **a** can contact **b** (directly or indirectly) or that $\mathbf{a}=\mathbf{b}$. It is easy to verify that **T** is a weak ordering of the set of individuals. It determines the equivalence relation $\mathbf{xTy} \wedge \mathbf{yTx}$, which may be read as "**x** and **y** can communicate with each other, or $\mathbf{x}=\mathbf{y}$."

This equivalence relation may be used to classify the individuals. Two men will be in the same equivalence class if they can communicate, that is, if each can contact the other one. The induced partial ordering $\mathbf{T}^*$ has a very intuitive meaning: The relation $\mathbf{uT^*v}$ holds if all members of the class **u** can contact all members of the class **v**, but not conversely unless $\mathbf{u}=\mathbf{v}$. Thus the partial ordering shows us the possible flow of information.

In particular, **u** is a maximal element of the partial ordering if its members cannot be contacted by members of any other class, and **u** is a minimal element if its members cannot contact members of other classes. Thus the maximal sets are message initiators, while the minimal sets are terminals for messages. (See M[4] Vol. II, Unit 2.)

It is interesting to study a given equivalence class. Any two members of such a class can communicate with each other. Hence any member can contact any other member. But how long does it take to contact other members? As a unit of time we will take the time needed to send a message from any one member to any member he can contact directly. We call this one *step.* We will assume that member *i* sends out a message, and we will be interested to know where the message could possibly be after *n* steps.

Let $\mathbf{N}_{ij}$ be the set of n such that a message starting from member i can be in member j's hands at the end of n steps. We will first consider $\mathbf{N}_{ii}$, the possible times at which a message can return to its originator. It is clear that if $a \in \mathbf{N}_{ii}$ and $b \in \mathbf{N}_{ii}$, then $a+b \in \mathbf{N}_{ii}$ after all the message can return in a steps and can be sent out again and be received back after b more steps. So the set $\mathbf{N}_{ii}$ is closed under addition. The following number-theoretic result will be useful. Its proof is given at the end of the section.

1.4.1 THEOREM. *A set of positive integers that is closed under addition contains all but a finite number of multiples of its greatest common divisor.*

If the greatest common divisor of the elements of $\mathbf{N}_{ii}$ is designated d_i, it is clear that the elements of $\mathbf{N}_{ii}$ are all multiples of d_i. But Theorem 1.4.1 tells us in addition that all sufficiently high multiples of d_i are in the set.

Since each member can contact every other member in its equivalence class, the $\mathbf{N}_{ij}$ are non-empty. We next prove that for i and j in the same equivalence class, $d_i = d_j = d$, and that the elements of a given $\mathbf{N}_{ij}$ are congruent to each other modulo d (their difference is a multiple of d). Suppose that $a \in \mathbf{N}_{ij}$, $b \in \mathbf{N}_{ij}$, and $c \in \mathbf{N}_{ji}$.

First of all, member i can contact himself by sending a message to member j and getting a message back. Hence $a+c \in \mathbf{N}_{ii}$. The message could also go to member j, come back to member j, and then go to member i. This could be done in $a+kd_j+c$ steps, where k is sufficiently large. Hence d_j must be a multiple of d_i. But in exactly the same way we can prove that d_i is a multiple of d_j. Hence $d_i = d_j = d$.

Or again, the message could go to member j in b steps, and then back to member i. Hence $b+c \in \mathbf{N}_{ii}$. Hence $a+c$ and $b+c$ are both divisible by d, and thus we see that $a \equiv b \pmod{d}$. Thus the elements of a given $\mathbf{N}_{ij}$ are congruent to each other modulo d. We can thus introduce numbers t_{ij}, with $0 \leqslant t_{ij} < d$, so that any element of $\mathbf{N}_{ij}$ is congruent to t_{ij}, modulo d. It is also easy to see that $\mathbf{N}_{ij}$ contains all but a finite number of the numbers $t_{ij}+kd$.

In particular we see that $t_{ii}=0$ in each case, and hence $t_{ij}+t_{ji} \equiv 0 \pmod{d}$. Also $t_{ij}+t_{jm} \equiv t_{im} \pmod{d}$. From this it is easily seen that $t_{ij}=0$ is an equivalence relation. Let us call such an equivalence class a *cyclic class*.

Since $t_{ij}+t_{jm} \equiv t_{im} \pmod{d}$, we see that $t_{ij}=t_{im}$ if and only if $t_{jm}=0$, hence if and only if members j and m are in the same cyclic class. Let n be any integer. If $n \equiv t_{ij} \pmod{d}$, then the message originating from member i can only be in this one cyclic class after n steps. From this it immediately follows that there are exactly d cyclic classes, and that the message moves cyclically from class to class, with cycle of length d. It is also easily seen that after sufficient time has elapsed, it can be in the hands of any member of the one cyclic class appropriate for n.

While this description of an equivalence class of the communication network holds in complete generality, the cycle degenerates when $d=1$. In this case there is a single "cyclic class," and after sufficient time has elapsed the message can be in the hands of any member at any time.

In particular, it is worth noting that if any member of the equivalence class can contact himself directly, then $d=1$. This is immediately seen from the fact that d is a divisor of any time in which a member can contact himself, and here d has to divide 1.

The number-theoretic result, § 1.4.1, is of such interest that its proof will be given here.

First of all we note that if the greatest common divisor d of the set is not 1, then we can divide all elements by d, and reduce the problem to the case $d=1$. Hence it suffices to treat this case. Here we have a set of numbers whose greatest common divisor is 1, and we must have a finite subset with this property. Hence, by a well-known result, there is a linear combination, $a_1n_1+a_2n_2+\cdots+a_kn_k$ of the elements (with positive or negative integers a_i) which is equal to 1. If we collect all the positive and all the negative terms separately, and remember that the set is closed under addition, we note that there must be elements m and n in the set, such that $m-n=1$ (m being the sum of the positive terms, and $-n$ the sum of the negative terms). Let q be any sufficiently large number, or more precisely $q \geqslant n(n-1)$. We can write $q=an+b$, where $a \geqslant (n-1)$ and $o \leqslant b \leqslant (n-1)$. Then we see that $q=(a-b)n+bm$, and hence q must be in the set.

§ **1.5 Probability measures.** In making a probability analysis of an experiment there are two basic steps. First, a set of logical possibilities is chosen. This problem was discussed in § 1.2. Second a probability measure is assigned. The way that this second step is carried out will be discussed in this section. We consider first a finite possibility space. (For a more detailed discussion see FM Chapter IV or FMS Chapter III.)

1.5.1 DEFINITION. *Let* $\mathbf{U}=\{\mathbf{a}_1, \mathbf{a}_2, \ldots, \mathbf{a}_r\}$ *be a set of logical possibilities. A* probability measure *for* $\mathbf{U}$ *is obtained by assigning to each element* $\mathbf{a}_j$ *a positive number* $\mathbf{w}(\mathbf{a}_j)$, *called a weight, in such a way that the weights assigned have sum* 1. *The* measure *of a subset* $\mathbf{A}$ *of* $\mathbf{U}$, *denoted by* $\mathbf{m}(A)$, *is the sum of the weights assigned to elements of* $\mathbf{A}$.

1.5.2 Theorem. *A probability measure* **m** *assigned to a possibility set* **U** *has the following properties:*

(1) For any subset **P** of **U**, $0 \leqslant \mathbf{m}(\mathbf{P}) \leqslant 1$.

(2) If **P** and **Q** are disjoint subsets of **U**, then $\mathbf{m}(\mathbf{P} \cup \mathbf{Q}) = \mathbf{m}(\mathbf{P}) + \mathbf{m}(\mathbf{Q})$.

(3) For any subsets **P** and **Q** of **U**, $\mathbf{m}(\mathbf{P} \cup \mathbf{Q}) = \mathbf{m}(\mathbf{P}) + \mathbf{m}(\mathbf{Q}) - \mathbf{m}(\mathbf{P} \cap \mathbf{Q})$.

(4) For any set **P** in **U**, $\mathbf{m}(\tilde{\mathbf{P}}) = 1 - \mathbf{m}(\mathbf{P})$.

1.5.3 Definition. *Let* **p** *be a statement relative to a set* **U** *having truth set* **P**. *The* probability of **p** relative to the probability measure **m** *is defined as* **m(P)**.

In any discussion where there is a fixed probability measure we shall refer simply to the probability of **p** without mentioning each time the measure. From Theorem 1.5.2 and the relation of the connectives to the set operations, we have the following theorem:

1.5.4 Theorem. *Let* **U** *be a set of possibilities for which a probability measure has been assigned. The probabilities of statements determined by this measure have the following properties:*

(1) *For any statement* **p**, $0 \leqslant \mathbf{Pr}[\mathbf{p}] \leqslant 1$.

(2) *If* **p** *and* **q** *are inconsistent then* $\mathbf{Pr}[\mathbf{p} \vee \mathbf{q}] = \mathbf{Pr}[\mathbf{p}] + \mathbf{Pr}[\mathbf{q}]$.

(3) *For any two statements* **p** *and* **q**, $\mathbf{Pr}[\mathbf{p} \vee \mathbf{q}] = \mathbf{Pr}[\mathbf{p}] + \mathbf{Pr}[\mathbf{q}] - \mathbf{Pr}[\mathbf{p} \wedge \mathbf{q}]$.

(4) *For any statement* **p**, $\mathbf{Pr}[\sim \mathbf{p}] = 1 - \mathbf{Pr}[\mathbf{p}]$.

1.5.5 Example. Given any finite set having s elements we can determine a probability measure by assigning weight $1/s$ to each element of **U**. This measure is called the *equiprobable* measure. For any set **A** with r elements, $\mathbf{m}(\mathbf{A}) = r/s$. For example, this is the measure which would normally be assigned to the outcomes for the roll of a die. In this case $\mathbf{U} = \{1, 2, 3, 4, 5, 6\}$ and a weight of $^1/_6$ is assigned to each.

1.5.6 Example. As an example of a situation where different weights would be assigned consider the following: A man observes a race between three horses **a**, **b**, and **c**. He feels that **a** and **b** have the same chance of winning but that **c** is twice as likely to win as **a**. We take the possibility set to be $\mathbf{U} = \{\mathbf{a}, \mathbf{b}, \mathbf{c}\}$ and assign weights $\mathbf{w}(\mathbf{a}) = {}^1/_4$, $\mathbf{w}(\mathbf{b}) = {}^1/_4$ and $\mathbf{w}(\mathbf{c}) = {}^1/_2$.

It is occasionally necessary to extend the above concepts to include the case of an experiment with an infinite sequence of possible outcomes. For example, consider the experiment of tossing a coin until the first time that a head turns up. The possible outcomes would be $\mathbf{U}=\{1, 2, 3, \ldots\}$. The above definitions and theorems apply equally well to this possibility set. We will have an infinite number of weights assigned but we still must require that they have sum 1. In the example just mentioned we would assign weights $(1/2, 1/4, 1/8, \ldots)$. These weights form a geometric progression having sum 1.

§ **1.6 Conditional probability.** It often happens that a probability measure has been assigned to a set **U** and then we learn that a certain statement **q** relative to **U** is true. With this new information we change the possibility set to the truth set **Q** of **q**. We wish to determine a probability measure on this new set from our original measure **m**. We do this by requiring that elements of **Q** should have the same relative weights as they had under the original assignment of weights. This means that our new weights must be the old weights multiplied by a constant to give them sum 1. This constant will be the reciprocal of the sum of the weights of all elements in **Q**, i.e. $1/\mathbf{m}(\mathbf{Q})$. (See FM Chapter IV or FMS Chapter III.)

1.6.1 DEFINITION. *Let* $\mathbf{U}=\{\mathbf{a}_1, \mathbf{a}_2, \ldots, \mathbf{a}_r\}$ *be a possibility set for which a measure has been assigned, determined by weights* $\mathbf{w}(\mathbf{a}_j)$. *Let* **q** *be a statement relative to* **U** (*not a self-contradiction*). *The* conditional probability measure given **q** *is a probability measure defined on* **Q** *the truth set of* **q**, *determined by weights*

$$\bar{\mathbf{w}}(\mathbf{a}_j) = \frac{\mathbf{w}(\mathbf{a}_j)}{\mathbf{m}(\mathbf{Q})}.$$

1.6.2 DEFINITION. *Let* **p** *and* **q** *be two statements relative to a set* **U** (**q** *not a self-contradiction*). *The conditional probability of* **p** *given* **q**, *denoted by* $\mathbf{Pr}[\mathbf{p}|\mathbf{q}]$ *is the probability of* **p** *computed from the conditional probability measure given* **q**.

1.6.3 THEOREM. *Let* **p** *and* **q** *be two statements relative to* **U** (**q** *not a self-contradiction*). *Assume that a probability measure* **m** *has been assigned to* **U**. *Then*

$$\mathbf{Pr}[\mathbf{p}|\mathbf{q}] = \frac{\mathbf{Pr}[\mathbf{p} \wedge \mathbf{q}]}{\mathbf{Pr}[\mathbf{q}]}$$

where $\mathbf{Pr}[\mathbf{p} \wedge \mathbf{q}]$ *and* $\mathbf{Pr}[\mathbf{q}]$ *are found from the measure* **m**.

1.6.4 EXAMPLE. In Example 1.5.6 assume that the man learns that horse **b** is not going to run. This causes him to consider the new possibility space $\mathbf{Q}=\{\mathbf{a}, \mathbf{c}\}$. The new weights which determine the conditional measure are $\bar{\mathbf{w}}(\mathbf{a})=\frac{1/4}{1/4+1/2}=1/3$ and $\bar{\mathbf{w}}(\mathbf{c})=\frac{1/2}{1/4+1/2}=2/3$.

We observe that it is still twice as likely that **c** will win than it is that **a** will win.

1.6.5 DEFINITION. *Two statements* **p** *and* **q** *(neither of which is a self-contradiction) are independent if* $\mathbf{Pr}[\mathbf{p} \wedge \mathbf{q}] = \mathbf{Pr}[\mathbf{p}] \cdot \mathbf{Pr}[\mathbf{q}]$.

It follows from Theorem 1.6.3 that **p** and **q** are independent if and only if $\mathbf{Pr}[\mathbf{p}|\mathbf{q}] = \mathbf{Pr}[\mathbf{p}]$ and $\mathbf{Pr}[\mathbf{q}|\mathbf{p}] = \mathbf{Pr}[\mathbf{q}]$. Thus to say that **p** and **q** are independent is to say that the knowledge that one is true does not effect the probability assigned to the other.

1.6.6 EXAMPLE. Consider two tosses of a coin. We describe the outcomes by $\mathbf{U} = \{\mathbf{HH}, \mathbf{HT}, \mathbf{TH}, \mathbf{TT}\}$. We assign the equiprobable measure. Let **p** be the statement "a head turns up on the first toss" and **q** the statement "a head turns up on the second toss." Then $\mathbf{Pr}[\mathbf{p} \wedge \mathbf{q}] = 1/4$, $\mathbf{Pr}[\mathbf{p}] = \mathbf{Pr}[\mathbf{q}] = 1/2$. Thus **p** and **q** are independent.

§ 1.7 Functions on a possibility space. Let $\mathbf{U} = \{\mathbf{a}_1, \mathbf{a}_2, \ldots, \mathbf{a}_r\}$ be a possibility space. Let **f** be a function with domain **U** and range $\mathbf{R} = \{\mathbf{r}_1, \mathbf{r}_2, \ldots, \mathbf{r}_s\}$. That is, **f** assigns to each element **U** a unique element of **R**. If **f** assigns $\mathbf{r}_k$ to $\mathbf{a}_j$, we write $\mathbf{f}(\mathbf{a}_j) = \mathbf{r}_k$. We write $\mathbf{f} = \mathbf{r}_k$ for the statement "the value of the function is $\mathbf{r}_k$." This is a statement relative to **U**, since its truth value is known when the outcome $\mathbf{a}_j$ is known. Hence it has a truth set which is a subset of **U**. (See FMS Chapters II, III, or M[4] Vol. II, Unit 1.)

1.7.1 DEFINITION. *Let* **f** *be a function with domain* **U** *and range* **R**. *Assume that a measure has been assigned to* **U**. *For each* $\mathbf{r}_k$ *in* **R** *let* $\mathbf{w}(\mathbf{r}_k) = \mathbf{Pr}[\mathbf{f} = \mathbf{r}_k]$. *The weights* $\mathbf{w}(\mathbf{r}_k)$ *determine a probability measure on the set* **R**, *called the* induced measure for **f**. *The weights are called the* induced weights.

We shall normally indicate the induced measure by giving both the range values and the weights in the form:

$$\mathbf{f}: \begin{Bmatrix} \mathbf{r}_1, & \mathbf{r}_2, & \ldots, & \mathbf{r}_s \\ \mathbf{w}(\mathbf{r}_1), & \mathbf{w}(\mathbf{r}_2), & \ldots, & \mathbf{w}(\mathbf{r}_s) \end{Bmatrix}.$$

Thus the induced weight of $\mathbf{r}_k$ in **R** is the measure of the truth set of $\mathbf{f} = \mathbf{r}_k$ in **U**.

1.7.2 EXAMPLE. In Example 1.6.6 let **f** be the function which gives the number of heads which turn up. The range of **f** is $\mathbf{R} = \{0, 1, 2\}$. The $\mathbf{Pr}[\mathbf{f} = 0] = 1/4$, $\mathbf{Pr}[\mathbf{f} = 1] = 1/2$, and $\mathbf{Pr}[\mathbf{f} = 2] = 1/4$. Hence the range and induced measure is:

$$\mathbf{f}: \begin{Bmatrix} 0 & 1 & 2 \\ 1/4 & 1/2 & 1/4 \end{Bmatrix}.$$

1.7.3 DEFINITION. *Let* **U** *be a possibility space, and* **f** *and* **g** *be two functions with domain* **U**, *each having as range a set of numbers. The function* $\mathbf{f}+\mathbf{g}$ *is the function with domain* **U** *which assigns to* $\mathbf{a}_j$ *the number* $\mathbf{f}(\mathbf{a}_j)+\mathbf{g}(\mathbf{a}_j)$. *The function* $\mathbf{f}\cdot\mathbf{g}$ *is the function with domain* **U** *which assigns to* $\mathbf{a}_j$ *the number* $\mathbf{f}(\mathbf{a}_j)\cdot\mathbf{g}(\mathbf{a}_j)$. *For any number* c *the* constant function **c** *is the function which assigns the number* c *to every element of* **U**.

Let **U** be a possibility space for which a measure has been assigned. Then if **f** and **g** are two numerical functions with domain **U**, $\mathbf{f}+\mathbf{g}$ and $\mathbf{f}\cdot\mathbf{g}$ will be functions with domain **U**, and as such have induced measures. In general there is no simple connection between the induced measures of these functions and the induced measure for **f** and **g**.

1.7.4 EXAMPLE. In Example 1.6.6 let **g** be a function having the value 1 if a head turns up on the first toss and 0 otherwise. Let **h** be a function having the value 1 if a head turns up on the second toss and 0 if a tail turns up. Then the range and induced measures for **g**, **h**, $\mathbf{g}+\mathbf{h}$, and $\mathbf{g}\cdot\mathbf{h}$ are

$$\mathbf{g}: \begin{Bmatrix} 0 & 1 \\ 1/2 & 1/2 \end{Bmatrix}$$

$$\mathbf{h}: \begin{Bmatrix} 0 & 1 \\ 1/2 & 1/2 \end{Bmatrix}$$

$$\mathbf{g}+\mathbf{h}: \begin{Bmatrix} 0 & 1 & 2 \\ 1/4 & 1/2 & 1/4 \end{Bmatrix}$$

$$\mathbf{g}\cdot\mathbf{h}: \begin{Bmatrix} 0 & 1 \\ 3/4 & 1/4 \end{Bmatrix}.$$

1.7.5 DEFINITION. *Let* **f** *be a function defined on* **U**. *Let* **p** *be a statement relative to* **U** *having truth set* **P**. *Assume that a measure* **m** *has been assigned to* **U**. *Let* $\mathbf{f}'$ *be the function* **f** *considered only on the set* **P**. *Then the induced measure for* $\mathbf{f}'$ *calculated from the conditional measure given* **p** *is called the* conditional induced measure for **f** given **p**.

1.7.6 DEFINITION. *Let* **f** *and* **g** *be two functions defined on a space* **U** *for which a probability measure has been assigned. Then* **f** *and* **g** *are* independent *if, for any* $\mathbf{r}_k$ *in the range of* **f** *and* $\mathbf{s}_j$ *in the range of* **g**, *the statements* $\mathbf{f}=\mathbf{r}_k$ *and* $\mathbf{g}=\mathbf{s}_j$ *are independent statements.*

An equivalent way to state the condition for independence of two functions is to say that the induced measure for one function is not changed by the knowledge of the value of the other.

§ 1.8 **Mean and variance of a function.** Throughout this section we shall assume that the functions considered are functions whose range set is a set of numbers. (A detailed discussion of the concepts introduced in this section is given in FMS Chapter III, or M[4] Vol. II, Unit 1.)

1.8.1 DEFINITION. *Let* $\mathbf{f}$ *be a function defined on a possibility space* $\mathbf{U}=\{\mathbf{a}_1, \mathbf{a}_2, \ldots, \mathbf{a}_r\}$, *for which a measure determined by weights* $\mathbf{w}(\mathbf{a}_j)$ *has been assigned. Then the* mean value *of* $\mathbf{f}$ *denoted by* $\mathbf{M}[\mathbf{f}]$ *is*

$$\mathbf{M}[\mathbf{f}]=\sum_j \mathbf{f}(\mathbf{a}_j)\cdot\mathbf{w}(\mathbf{a}_j).$$

The term expected value *is often used in place of mean value.*

1.8.2 THEOREM. *Let* $\mathbf{f}$ *be a function defined on* $\mathbf{U}$. *Assume that for a probability measure* $\mathbf{m}$ *defined on* $\mathbf{U}$, *the function* $\mathbf{f}$ *has induced measure*

$$\mathbf{f}: \begin{Bmatrix} \mathbf{r}_1, & \mathbf{r}_2, & \ldots, & \mathbf{r}_s \\ \mathbf{w}(\mathbf{r}_1), & \mathbf{w}(\mathbf{r}_2), & \ldots, & \mathbf{w}(\mathbf{r}_s) \end{Bmatrix}.$$

Then

$$\mathbf{M}[\mathbf{f}]=\sum_j \mathbf{r}_j\cdot\mathbf{w}(\mathbf{r}_j).$$

1.8.3 EXAMPLE. In Example 1.6.6 let $\mathbf{f}$ be the number of heads which turn up. From the definition of mean value we have

$$\begin{aligned}\mathbf{M}[\mathbf{f}] &= \mathbf{f}(\mathbf{HH})\cdot{}^1/_4+\mathbf{f}(\mathbf{HT})\cdot{}^1/_4+\mathbf{f}(\mathbf{TH})\cdot{}^1/_4+\mathbf{f}(\mathbf{TT})\cdot{}^1/_4 \\ &= 2\cdot{}^1/_4+1\cdot{}^1/_4+1\cdot{}^1/_4+0\cdot{}^1/_4 \\ &= 1.\end{aligned}$$

We can also calculate the mean of $\mathbf{f}$ by making use of Theorem 1.8.2. The range and induced measure for $\mathbf{f}$ is

$$\mathbf{f}: \begin{Bmatrix} 0 & 1 & 2 \\ {}^1/_4 & {}^1/_2 & {}^1/_4 \end{Bmatrix}.$$

Thus by Theorem 1.8.2,

$$\mathbf{M}[\mathbf{f}] = 0\cdot{}^1/_4+1\cdot{}^1/_2+2\cdot{}^1/_4 = 1.$$

1.8.4 DEFINITION. *Let* $\mathbf{f}$ *be a function defined on a possibility space* $\mathbf{U}$ *for which a measure has been assigned. Let* $\mathbf{M}[\mathbf{f}]=m$ *be the mean of this function. Then the* variance *of* $\mathbf{f}$, *denoted by* $\mathbf{Var}[\mathbf{f}]$, *is the mean of the function* $(\mathbf{f}-m)^2$. *The* standard deviation *denoted by* $\mathbf{sd}[\mathbf{f}]$, *is the square root of the variance.*

1.8.5 THEOREM. *Let* **f** *be a function having mean value* m. *Then* $\mathbf{Var[f]} = \mathbf{M[f^2]} - m^2$.

1.8.6 EXAMPLE. Let **f** be the function in Example 1.8.3. We found that $\mathbf{M[f]} = 1$. Thus

$$\begin{aligned}\mathbf{Var[f]} &= (2-1)^2 \cdot {}^1/_4 + (1-1)^2 \cdot {}^1/_4 + (1-1)^2 \cdot {}^1/_4 + (0-1)^2 \cdot {}^1/_4 \\ &= {}^1/_2.\end{aligned}$$

An alternative way to compute the variance is to make use of Theorem 1.8.5. Using this result we find

$$\mathbf{M[f^2]} = 4 \cdot {}^1/_4 + 1 \cdot {}^1/_4 + 1 \cdot {}^1/_4 + 0 \cdot {}^1/_4 = {}^3/_2.$$

Since $\mathbf{M[f]} = 1$, we have $\mathbf{Var[f]} = {}^3/_2 - 1 = {}^1/_2$.

1.8.7 THEOREM. *Let* **f** *and* **g** *be any two functions for which means and variances have been defined. Then*

(1) $\mathbf{M[c]} = c$.
(2) $\mathbf{M[f+g]} = \mathbf{M[f]} + \mathbf{M[g]}$.
(3) $\mathbf{M[c \cdot f]} = \mathbf{c \cdot M[f]}$.
(4) $\mathbf{Var[c \cdot f]} = c^2 \cdot \mathbf{Var[f]}$.
(5) $\mathbf{Var[c+f]} = \mathbf{Var[f]}$.
(6) $\mathbf{Var[c]} = 0$.

If **f** *and* **g** *are independent functions then*

(7) $\mathbf{M[f \cdot g]} = \mathbf{M[f] \cdot M[g]}$.
(8) $\mathbf{Var[f+g]} = \mathbf{Var[f]} + \mathbf{Var[g]}$.

1.8.8 DEFINITION. *Let* **p** *be a statement relative to a possibility set* **U** *for which a measure has been assigned. Let* **f** *be a function with domain* **U**. *The* conditional mean *and* variance of **f** given **p** *are the mean and variance of* **f** *computed from the conditional measure given* **p**. *We denote these by* $\mathbf{M[f|p]}$ *and* $\mathbf{Var[f|p]}$.

1.8.9 THEOREM. *Let* $\mathbf{p}_1, \mathbf{p}_2, \ldots, \mathbf{p}_r$ *be a complete set of alternatives relative to a set* **U**. *Let* **f** *be a function with domain* **U**. *Then*

$$\mathbf{M[f]} = \sum_j \mathbf{M[f|p_j] \cdot Pr[p_j]}.$$

1.8.10 THEOREM. *If* $\mathbf{f}_1, \mathbf{f}_2, \ldots$ *is a sequence of functions such that for some constant* c,

$$\mathbf{M}[(\mathbf{f}_n - c)^2] \to 0$$

as $n \to \infty$, *then*

$$\mathbf{M[f}_n] \to c$$

and for any $\epsilon > 0$

$$\mathbf{Pr}[|\mathbf{f}_n - c| > \epsilon] \to 0$$

as $n \to \infty$.

1.8.11 DEFINITION. *Let* $\mathbf{f}_1$ *and* $\mathbf{f}_2$ *be two functions with* $\mathbf{M}[\mathbf{f}_i]=a_i$ *and* $\mathbf{sd}[\mathbf{f}_i]=b_i$. *Then the* covariance *of* $\mathbf{f}_1$ *and* $\mathbf{f}_2$ *is defined by*

$$\mathbf{Cov}[\mathbf{f}_1, \mathbf{f}_2] = \mathbf{M}[(\mathbf{f}_1-a_1)(\mathbf{f}_2-a_2)],$$

and the correlation *of* $\mathbf{f}_1$ *and* $\mathbf{f}_2$ *is*

$$\mathbf{Corr}[\mathbf{f}_1, \mathbf{f}_2] = \frac{\mathbf{Cov}[\mathbf{f}_1, \mathbf{f}_2]}{b_1 \cdot b_2},$$

§ **1.9 Stochastic processes.** In this section we shall briefly describe the concept of a stochastic process. A more complete treatment may be found in FM Chapter IV or FMS Chapter III.

We wish to give a probability measure to describe an experiment which takes place in stages. The outcome at the n-th stage is allowed to depend on the outcomes of the previous stages. It is assumed, however, that the probability for each possible outcome at a particular stage is known when the outcomes of all previous stages are known. From this knowledge we shall construct a possibility space and measure for the over-all experiment.

We shall illustrate the construction of the possibility space and measure by a particular example. The general procedure will be clear from this.

1.9.1 EXAMPLE. We choose at random one of two coins **A** or **B**. Coin **A** is a fair coin and coin **B** has heads on both sides. The coin chosen is tossed. If a tail comes up a die is rolled. If a head turns up the coin is thrown again. The first stage of the experiment is the choice of a coin. At the second stage, a coin is tossed. At the third stage a coin is tossed or a die is rolled, depending on the outcome of the first two stages.

We indicate the possible outcomes of the experiment by a *tree* as shown in Figure 1-1.

The possibilities for the experiment are $\mathbf{t}_1=(\mathbf{A}, \mathbf{H}, \mathbf{H})$, $\mathbf{t}_2=(\mathbf{A}, \mathbf{H}, \mathbf{T})$, $\mathbf{t}_3=(\mathbf{A}, \mathbf{T}, 1)$, $\mathbf{t}_4=(\mathbf{A}, \mathbf{T}, 2)$, etc. Each possibility may be identified with a *path* through the trees. Each path is made up of line segments called *branches*. In the tree we have just given, there are nine paths each having three branches.

We know the probability for each outcome at a given stage when the previous stages are known. For example, if outcome **A** occurs on the first stage and **T** on the second stage, then the probability of a 1 for the third stage is $^1/_6$. We assign these known probabilities to the branches and call them *branch probabilities*.

We next assign weights to the paths equal to the product of the probabilities assigned to the components of the path. For example the path t_7 corresponds to outcome **A** on the first stage, **T** on the second, and 5 on the third. The weight assigned to this path is

$$^1/_2 \cdot {}^1/_2 \cdot {}^1/_6 = {}^1/_{24}.$$

This procedure assigns a weight to each path of the tree and the sum of the weights assigned is 1. The set **U** of all paths may be considered a suitable possibility space for the consideration of any statement whose truth value depends on the outcome of the total experiment. The measure assigned by the path weights is the appropriate probability measure.

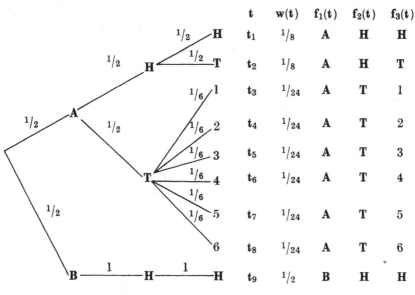

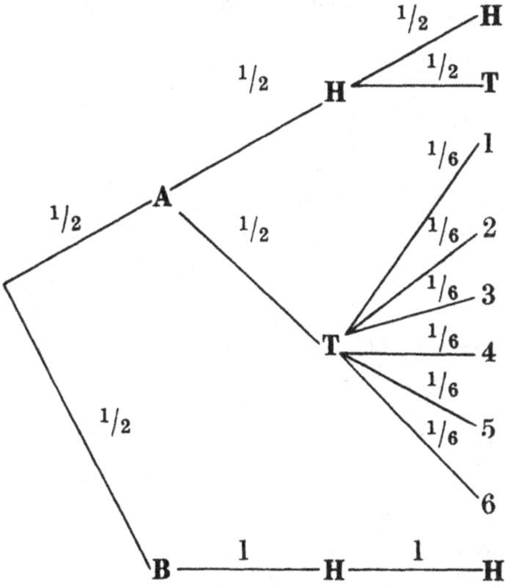

t	w(t)	f_1(t)	f_2(t)	f_3(t)
t_1	$^1/_8$	A	H	H
t_2	$^1/_8$	A	H	T
t_3	$^1/_{24}$	A	T	1
t_4	$^1/_{24}$	A	T	2
t_5	$^1/_{24}$	A	T	3
t_6	$^1/_{24}$	A	T	4
t_7	$^1/_{24}$	A	T	5
t_8	$^1/_{24}$	A	T	6
t_9	$^1/_2$	B	H	H

FIGURE 1-1

The above procedure can be carried out for any experiment that takes place in stages. We require only that there be a finite number of possible outcomes at each stage and that we know the probabilities for any particular outcome at the j-th stage, given the knowledge of the outcome for the first $j-1$ stages. For each j we obtain a tree $\mathbf{U}_j$. The set of paths of this tree serves as a possibility space for any statement relating to the first j experiments. On this tree we assign a measure to the set of all paths. We first assign branch probabilities. Then the weight assigned to a path is the product of all branch probabilities on the path. The tree measures are consistent in the following sense. A statement whose truth value depends only on the first j stages may be considered a statement relative to any tree $\mathbf{U}_i$ for $i \geqslant j$.

Each of these trees has its own tree measure and the probability of the statement could be found from any one of these measures. However, in every case the same probability would be assigned.

Assume that we have a tree for an n stage experiment. Let $\mathbf{f}_j$ be a function with domain the set of paths $\mathbf{U}_n$ and value the outcome at the j-th stage. Then the functions $\mathbf{f}_1, \mathbf{f}_2, \ldots, \mathbf{f}_n$ are called *outcome functions.* The set of functions $\mathbf{f}_1, \mathbf{f}_2, \ldots, \mathbf{f}_n$ is called a *stochastic process.* (In Markov chain theory it is convenient to denote the first outcome by $\mathbf{f}_0$ instead of $\mathbf{f}_1$.)

In our example there are three outcome functions. We have indicated in Figure 1-1 the value of each function on each path.

There is a simple connection between the branch probabilities and the outcome functions. The branch probabilities at the first stage are,

$$\mathbf{Pr}[\mathbf{f}_1 = \mathbf{r}_i]$$

at the second stage

$$\mathbf{Pr}[\mathbf{f}_2 = \mathbf{r}_j | \mathbf{f}_1 = \mathbf{r}_i]$$

at the third stage

$$\mathbf{Pr}[\mathbf{f}_3 = \mathbf{r}_k | \mathbf{f}_2 = \mathbf{r}_j \wedge \mathbf{f}_1 = \mathbf{r}_i]$$

etc.

In our example,

$$\mathbf{Pr}[\mathbf{f}_1 = \mathbf{A}] = \mathbf{w}(\mathbf{t}_1) + \cdots + \mathbf{w}(\mathbf{t}_8) = {}^1/_2$$

$$\mathbf{Pr}[\mathbf{f}_2 = \mathbf{T} | \mathbf{f}_1 = \mathbf{A}] = \frac{\mathbf{Pr}[\mathbf{f}_2 = \mathbf{T} \wedge \mathbf{f}_1 = \mathbf{A}]}{\mathbf{Pr}[\mathbf{f}_1 = \mathbf{A}]}$$

$$= \frac{\mathbf{w}(\mathbf{t}_3) + \cdots + \mathbf{w}(\mathbf{t}_8)}{\mathbf{w}(\mathbf{t}_1) + \cdots + \mathbf{w}(\mathbf{t}_8)} = \frac{{}^1/_4}{{}^1/_2} = {}^1/_2$$

$$\mathbf{Pr}[\mathbf{f}_3 = 1 | \mathbf{f}_2 = \mathbf{T} \wedge \mathbf{f}_1 = \mathbf{A}] = \frac{\mathbf{Pr}[\mathbf{f}_3 = 1 \wedge \mathbf{f}_2 = \mathbf{T} \wedge \mathbf{f}_1 = \mathbf{A}]}{\mathbf{Pr}[\mathbf{f}_2 = \mathbf{T} \wedge \mathbf{f}_1 = \mathbf{A}]}$$

$$= \frac{\mathbf{w}(\mathbf{t}_3)}{\mathbf{w}(\mathbf{t}_3) + \cdots + \mathbf{w}(\mathbf{t}_8)} = \frac{{}^1/_{24}}{{}^1/_4} = {}^1/_6.$$

1.9.2 EXAMPLE. We shall often deal with experiments where we allow an arbitrary number of stages. For example, in considering the tosses of a coin, we can envision any number of tosses. The tree for three tosses and the path measure is shown in Fig. 1.2.

For any number of tosses we can construct a tree. It is even possible to consider continuing the tree indefinitely to obtain a tree with infinite paths. Our procedure for assigning a measure would not in this case be adequate since it would assign weight 0 to every path. We shall not, however, have to assign a measure to the infinite tree. This is the case because the statements about the process that interest us will depend only on a finite part of the tree, and for any finite number of stages we have a method of assigning a measure. We shall, however, consider functions whose definition requires the infinite tree.

For example, in Example 1.9.2 let the value of f be the stage at which the first head occurs. Then f is defined for all paths with at least one head. This is a subset of paths in the infinite tree. We shall

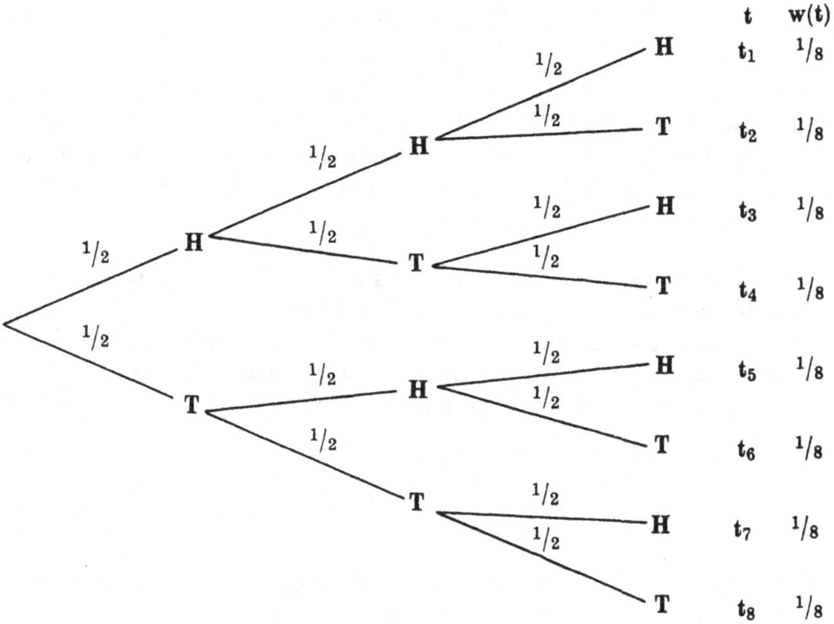

FIGURE 1-2

speak of the mean value of such a function when the following conditions are satisfied:

(a) There is a sequence of numerical range values $\mathbf{r}_1, \mathbf{r}_2, \ldots$ such that the truth value of the statement $\mathbf{f}=\mathbf{r}_j$ depends only on the outcomes of a finite number of stages and $\sum_j \mathbf{Pr}[\mathbf{f}=\mathbf{r}_j] = 1$.

(b) $\sum_j \mathbf{r}_j\mathbf{Pr}[\mathbf{f}=\mathbf{r}_j] < \infty$.

In case (a) and (b) hold, we say that f has a mean value given by

$$\mathbf{M}[\mathbf{f}] = \sum_j \mathbf{r}_j\mathbf{Pr}[\mathbf{f} = \mathbf{r}_j]$$

When f has a mean a, we shall say that f has a variance if $(\mathbf{f}-a)^2$ has a mean. If so, $\mathbf{Var}[\mathbf{f}]=\mathbf{M}[(\mathbf{f}-a)^2]$.

All properties of means and variances given in § 8 hold for these extended mean values. In addition we shall need the following theorem.

1.9.3 THEOREM. *Let* $\mathbf{f}_1, \mathbf{f}_2, \ldots$ *be functions such that the range of each* $\mathbf{f}_j$ *is a subset of the same finite set of numbers. Let* $\mathbf{s} = \mathbf{f}_1 + \mathbf{f}_2 + \cdots$. *Then if the mean of* $\mathbf{s}$ *exists,*

$$\mathbf{M}[\mathbf{s}] = \sum_j \mathbf{M}[\mathbf{f}_j].$$

A stochastic process for which the outcome functions all have ranges which are subsets of a given finite set is called a *finite stochastic process.* Thus Theorem 1.9.3 states that in a finite stochastic process the mean of the sum of the functions (if this mean exists) is the sum of the means of the functions.

§ 1.10 Summability of sequences and series. It may occur that for a divergent sequence $s_0, s_1, s_2, \ldots$ we can form a sequence of averages of the terms, and that this new sequence converges. In this case we say that the original sequence is *summable* by means of the averaging process. We will be concerned with only two methods of averaging.

Let $t_n = (1/n)\sum_{i=0}^{n-1} s_i$ and let $u_n = \sum_{i=0}^{n} \binom{n}{i} k^{n-i}(1-k)^i s_i$ for some k such that $0 < k < 1$. Each of these is an average of terms of the sequence, with non-negative coefficients whose sum is 1. If the sequence $t_1, t_2, \ldots$ converges to a limit t, then we say that the original sequence is *Cesaro-summable* to t. If the sequence $u_1, u_2, \ldots$ converges to u, then we say that the original sequence is *Euler-summable* to u.

For example, consider the sequence $1, 0, 1, 0, 1, 0, \ldots$. We find that $t_n = \frac{1}{2}$ if n is even, $\frac{1}{2} + \frac{1}{2n}$ if n is odd. This sequence converges to $\frac{1}{2}$, and hence the original sequence is Cesaro-summable to $1/2$. It is easy to verify that $\lim_{n\to\infty} u_n = 1/2$ and hence the original sequence is also Euler-summable to $1/2$. But the original sequence diverges.

These two summability methods have the following two properties: (1) If a sequence converges, then it is summable by each method to its limit. (2) If a sequence is summable by both methods, the two sums must be the same.

Summability may also be applied to a series. To say that the series $\sum_{k=0}^{\infty} a_k$ is summable by a given method means that its sequence of partial sums $s_i = \sum_{k=0}^{i} a_k$ is summable by that method. For example if we apply Cesaro-summability to the partial sums, we obtain $t_n = \sum_{k=0}^{n-1} \frac{n-k}{n} a_k$.

§ **1.11 Matrices.** A matrix is a rectangular array of numbers. An $r \times s$ matrix has r rows and s columns, a total of rs entries (or components). Three special kinds of matrices will be especially important in this book. A matrix having the same number of rows as columns is called a *square matrix*. That is, a square matrix is $r \times r$. If $r = 1$, that is, the matrix consists of a single row, then we call it a *row vector*. If $s = 1$, i.e. the matrix has a single column, we call it a column vector. Matrices will be denoted by capitals and vectors by small Greek letters.

Let the $r \times s$ matrix A have components a_{ij}, and the $r' \times s'$ matrix B have components b_{ij}. Then we define the following operations and relations:

(1) The matrix kA has components ka_{ij}. That is, a multiplication of the matrix by a number means multiplying each component by this number. The matrix $-A$ is $(-1)A$.

(2) If $r = r'$ and $s = s'$, then the matrix sum $A + B$ has components $a_{ij} + b_{ij}$. That is, addition is carried out componentwise.

(3) If $s = r'$, we define the product AB to have components $\sum_{k=1}^{s} a_{ik} b_{kj}$. Note that the product of an $r \times s$ and $s \times t$ matrix is an $r \times t$ matrix. This definition also applies to the product of a row vector and a matrix, αA, or to a matrix times a column vector, $A\beta$. In the former case the product of a $1 \times r$ and an $r \times s$ matrix is a $1 \times s$ matrix, or a row vector. If the matrix A is square, the resulting row vector has the same number of components as α. Thus a square matrix may be thought of as a transformation of row vectors. Similarly we can think of it as a transformation of column vectors. This will be our principal use of the product of a vector and a matrix.

(4) We say that $A \geqslant B$ (or that $A = B$) if $a_{ij} \geqslant b_{ij}$ (or $a_{ij} = b_{ij}$) for all i and j. That is, matrix relations must hold componentwise—for all corresponding components.

(5) Some special matrices play an important role. The $r \times s$ matrix having all components equal to 0 is denoted by $O_{r \times s}$. The subscripts are omitted whenever there is no danger of confusion. The $r \times r$ matrix having 1's as components a_{ii} ("on

the main diagonal") and 0's elsewhere is denoted by I_r. The subscript is often omitted. The role that these matrices play can be seen as follows. Let A, I, and O be $r \times r$, let α be an r-component row vector, and β an r-component column vector. Then:

$$\begin{aligned}
A+O &= O+A = A \\
A+(-A) &= (-A)+A = O \\
AI &= IA = A \\
\alpha I &= \alpha \\
I\beta &= \beta \\
AO &= OA = O \\
O\beta &= 0 \\
\alpha O &= 0.
\end{aligned}$$

Thus the matrices O and I play somewhat the same role as the numbers 0 and 1.

(6) In analogy to the reciprocal of a number we define the *inverse* of a matrix. The $r \times r$ matrix B is said to be the inverse of the $r \times r$ matrix A if $AB = I$. If such an inverse exists, it is denoted by A^{-1}. The inverse can be found by solving r^2 simultaneous equations. Of course, these equations may fail to have a solution. But when they do have a solution, the solution is unique, and we can show that $AA^{-1} = A^{-1}A = I$.

The various arithmetical operations on matrices, whenever they are defined, obey the usual laws of arithmetic. The one major exception to this is that matrix multiplication is not commutative, i.e. that AB need not equal BA. One important case where matrices commute is the case of powers of a given matrix. Let A^n be A multiplied by itself n times. Then $A^n \cdot A^m = A^m \cdot A^n$ for every n and m. We define $A^0 = I$.

It is convenient to introduce row vector η_r and the column vector ξ_r having all components equal to 1. The subscript is again omitted when possible. These vectors are convenient for summing vectors or rows and columns of matrices. The product $\alpha\xi$ is a number (or more precisely a matrix with a single entry) which is the sum of the components of α. Similarly for $\eta\beta$. The product $A\xi$ is a column vector whose i-th component gives the sum of the components in the i-th row of A (or the i-th row sum of A). Similarly ηA gives the column sums of A. We shall denote by E a square matrix with all entries 1. Note that $E = \xi\eta$.

Let us give some examples of these operations and relations.

$$3\begin{pmatrix}2 & 1\\ 0 & -1\end{pmatrix} = \begin{pmatrix}6 & 3\\ 0 & -3\end{pmatrix}$$

$$\begin{pmatrix}2 & 1\\ 0 & -1\end{pmatrix} + \begin{pmatrix}-1 & 0\\ 0 & -2\end{pmatrix} = \begin{pmatrix}1 & 1\\ 0 & -3\end{pmatrix}$$

$$(1, 2, 3) + (2, 1, 0) = (3, 3, 3)$$

$$(1, 2, 3)\begin{pmatrix}2 & 1\\ 0 & -1\\ 1 & 0\end{pmatrix} = (5, -1)$$

$$\begin{pmatrix}2 & 1 & 0\\ 0 & 1 & 2\end{pmatrix}\begin{pmatrix}2 & 1\\ 0 & -1\\ 1 & 0\end{pmatrix} = \begin{pmatrix}4 & 1\\ 2 & -1\end{pmatrix}$$

$$\begin{pmatrix}2 & 1\\ 0 & -1\end{pmatrix}\begin{pmatrix}1\\ 1\end{pmatrix} = \begin{pmatrix}3\\ -1\end{pmatrix}$$

$$\begin{pmatrix}1\\ 0\\ -1\end{pmatrix} \geqslant \begin{pmatrix}1/2\\ 0\\ -2\end{pmatrix}$$

$$\begin{pmatrix}2 & 1\\ 1 & 1\end{pmatrix}\begin{pmatrix}1 & -1\\ -1 & 2\end{pmatrix} = \begin{pmatrix}1 & -1\\ -1 & 2\end{pmatrix}\begin{pmatrix}2 & 1\\ 1 & 1\end{pmatrix} = \begin{pmatrix}1 & 0\\ 0 & 1\end{pmatrix} = I.$$

Therefore,

$$\begin{pmatrix}1 & -1\\ -1 & 2\end{pmatrix} = \begin{pmatrix}2 & 1\\ 1 & 1\end{pmatrix}^{-1}.$$

For a square matrix A we introduce its *transpose* A^T. The ij-th entry of A^T is the ji-th entry of A. We also define the matrix A_{dg} which agrees with A on the main diagonal, but is 0 elsewhere. The matrix A_{sq} is formed from A by squaring each entry. This, of course, will not normally be the same as A^2. (But $D^2 = D_{sq}$ for a *diagonal matrix* D, i.e. a matrix whose only non-zero entries are on the main diagonal.) Similarly we define α_{sq} for a vector α.

It is often convenient to give a matrix or a vector in terms of its components. We thus write $\{a_{ij}\}$ for the matrix whose ij component is a_{ij}. Similarly we write $\{a_j\}$ for a row-vector, and $\{a_i\}$ for a column vector. The following relations will illustrate this notation.

$$\begin{aligned}\{a_{ij}\}+\{b_{ij}\} &= \{a_{ij}+b_{ij}\},\\ O &= \{0\},\\ \xi\eta = E &= \{1\},\\ \{a_{ij}\}_{\text{sq}} &= \{a^2{}_{ij}\},\\ \{a_{ij}\}^T &= \{a_{ji}\},\\ 3\{a_i\} &= \{3a_i\},\\ \{a_i\}\{b_j\} &= \{a_ib_j\}.\end{aligned}$$

The last example shows that the product of a column vector and a row vector (each with r components) is a matrix (with $r \times r$ components). This must be contrasted with the product in the reverse order, which is a single component. For example, if α is a row vector, then $\alpha\xi$ gives the sum of its components. However, $\xi\alpha$ gives an $r \times r$ matrix with α for each row.

Suppose that we have a sequence of matrices A_k, with entries $a^{(k)}{}_{ij}$. We will say that the series $A_0+A_1+A_2+\cdots$ converges if each series of entries converges, i.e. if $a^{(0)}{}_{ij}+a^{(1)}{}_{ij}+a^{(2)}{}_{ij}+\cdots$ converges for every i and j. And if the sum of this series of components is a_{ij}, for each i and j, and if A is the matrix with these entries as components, then we say that A is the sum of the infinite series of matrices. In brief, we define an infinite sum of matrices by forming the sum for each component.

1.11.1 THEOREM. *If A^n tends to O (zero matrix) as n tends to infinity, then $(I-A)$ has an inverse, and*

$$(I-A)^{-1} = I+A+A^2+A^3+\cdots = \sum_{k=0}^{\infty} A^k.$$

PROOF. Consider the identity

$$(I-A)\cdot(I+A+A^2+\cdots+A^{n-1}) = I-A^n,$$

which is easily verified by multiplying out the left side. By hypothesis we know that the right side tends to I. This matrix has determinant 1. Hence for sufficiently large n, $I-A^n$ must have a non-zero determinant. But the determinant of a product of two matrices is the product of the determinants, hence $I-A$ cannot have a zero determinant. The determinant not being equal to zero is a sufficient condition for a matrix to have an inverse. Hence $I-A$ has an inverse. Since this inverse exists, we may multiply both sides of the identity by it:

$$I+A+A^2+\cdots+A^{n-1} = (I-A)^{-1}\cdot(I-A^n).$$

But the right side of this new identity clearly tends to $(I-A)^{-1}$, which completes the proof.

One can define the summability of matrix sequences and series exactly as in § 1.10, applying the averaging method to each component of the matrix. Then there is a generalization of the previous theorem: If the sequence A^n is summable to O by some averaging method, then the matrix $I-A$ has an inverse, and the series $I+A+A^2+\cdots$ is summable by the same method to $(I-A)^{-1}$.

1.11.2 DEFINITION. *A square matrix A is positive semi-definite if for any column vector γ, $\gamma^T A\gamma \geqslant 0$.*

1.11.3 THEOREM. *For any positive semi-definite matrix A there is a matrix B such that $A = B^T B$.*

CHAPTER II

BASIC CONCEPTS OF MARKOV CHAINS

§ 2.1 Definition of a Markov process and a Markov chain. We recall that for a finite stochastic process we have a tree and a tree measure and a sequence of outcome functions $\mathbf{f}_n$, $n=0, 1, 2, \ldots$. The domain of $\mathbf{f}_n$ is the tree $\mathbf{T}_n$ and the range is the set $\mathbf{U}_n$ of possible outcomes for the n-th experiment. The value of $\mathbf{f}_n$ is $\mathbf{s}_j$ if the outcome of the n-th experiment is $\mathbf{s}_j$ (see § 1.9). In the following definitions, whenever a conditional probability $\mathbf{Pr}[\mathbf{q}|\mathbf{p}]$ occurs, it is assumed that $\mathbf{p}$ is not logically false. The reader may find it convenient from time to time to refer to the summary of basic notations and quantities at the end of the book.

A finite stochastic process is an *independent process* if

(I) *For any statement* $\mathbf{p}$ *whose truth value depends only on the outcomes before the n-th,*

$$\mathbf{Pr}[\mathbf{f}_n=\mathbf{s}_j|\mathbf{p}] = \mathbf{Pr}[\mathbf{f}_n=\mathbf{s}_j].$$

For such a process the knowledge of the outcome of any preceding experiment does not affect our predictions for the next experiment. For a Markov process we weaken this to allow the knowledge of the immediate past to influence these predictions.

2.1.1 DEFINITION. *A* finite Markov process *is a finite stochastic process such that*

(II) *For any statement* $\mathbf{p}$ *whose truth value depends only on the outcomes before the n-st,*

$$\mathbf{Pr}[\mathbf{f}_n=\mathbf{s}_j|(\mathbf{f}_{n-1}=\mathbf{s}_i)\wedge\mathbf{p}] = \mathbf{Pr}[\mathbf{f}_n=\mathbf{s}_j|\mathbf{f}_{n-1}=\mathbf{s}_i].$$

We shall refer to condition II as the *Markov property.* For a Markov process, knowing the outcome of the last experiment we can neglect any other information we have about the past in predicting the future. It is important to realize that this is the case only if we know exactly the outcome of the last experiment. For example, if we know only that the outcome of the last experiment was either $\mathbf{s}_i$ or $\mathbf{s}_k$ then knowledge of the truth value of a statement $\mathbf{p}$ relating to earlier experiments may affect our future predictions.

2.1.2 DEFINITION. *The* n-*th* step transition probabilities *for a Markov process, denoted by $p_{ij}(n)$ are*

$$p_{ij}(n) = \mathbf{Pr}[\mathbf{f}_n = \mathbf{s}_j | \mathbf{f}_{n-1} = \mathbf{s}_i].$$

2.1.3 DEFINITION. *A* finite Markov chain *is a finite Markov process such that the transition probabilities $p_{ij}(n)$ do not depend on n. In this case they are denoted by p_{ij}. The elements of* $\mathbf{U}$ *are called* states.

2.1.4 DEFINITION. *The* transition matrix *for a Markov chain is the matrix P with entries p_{ij}. The* initial probability vector *is the vector $\pi_0 = \{p_j^{(0)}\} = \{\mathbf{Pr}[\mathbf{f}_0 = \mathbf{s}_j]\}$.*

For a Markov chain we may visualize a process which moves from state to state. It starts in $\mathbf{s}_j$ with probability $p^{(0)}{}_j$. If at any time it is in state $\mathbf{s}_i$, then it moves on the next "step" to $\mathbf{s}_j$ with probability p_{ij}. The initial probabilities are thought of as giving the probabilities for the various possible starting states. The initial probability vector and the transition matrix completely determine the Markov chain process, since they are sufficient to build the entire tree measure. Thus, given any probability vector π_0 and any probability matrix P, there is a unique Markov chain (except possibly for renaming the states) which will have the π_0 as initial probability vector and P as transition matrix.

In most of our discussions we will consider a fixed transition matrix P, but we will wish to vary the initial vector π. The tree measure assigned will depend on the initial vector π that is chosen. Hence if $\mathbf{p}$ is any statement relative to the tree, or $\mathbf{f}$ is a function with domain the tree, $\mathbf{Pr}[\mathbf{p}]$, $\mathbf{M}[\mathbf{f}]$, and $\mathbf{Var}[\mathbf{f}]$ all depend on π. We indicate this by writing $\mathbf{Pr}_\pi[\mathbf{p}]$, $\mathbf{M}_\pi[\mathbf{f}]$ and $\mathbf{Var}_\pi[\mathbf{f}]$. The special case where π has a 1 in the i-th component (process is started in state $\mathbf{s}_i$) is denoted $\mathbf{Pr}_i[\mathbf{p}]$, $\mathbf{M}_i[\mathbf{f}]$, $\mathbf{Var}_i[\mathbf{f}]$.

We shall give several examples of Markov chains in the next section. We conclude this section with a few brief remarks about the Markov property.

It can be easily proved that the Markov property is equivalent to the following property more symmetric with respect to time.

(II′) *Let* $\mathbf{p}$ *be any statement whose truth value depends only on outcomes after the n-th experiment and* $\mathbf{q}$ *be any statement whose truth value depends only on outcomes before the n-th experiment. Then*

$$\mathbf{Pr}[\mathbf{p} \wedge \mathbf{q} | \mathbf{f}_n = \mathbf{s}_j] = \mathbf{Pr}[\mathbf{p} | \mathbf{f}_n = \mathbf{s}_j] \cdot \mathbf{Pr}[\mathbf{q} | \mathbf{f}_n = \mathbf{s}_j].$$

This condition says essentially that, given the present, the past and future are independent of each other. This more symmetric definition suggests in turn that a Markov process should remain a Markov process if it is observed in reverse order. That the latter is true is seen from the following theorem. (We shall not prove this theorem.)

2.1.5 THEOREM. *Given a Markov process let* $\mathbf{p}$ *be any statement whose truth value depends only on experiments after the n-th experiment. Then*

$$\mathbf{Pr}[\mathbf{f}_n = \mathbf{s}_j | (\mathbf{f}_{n+1} = \mathbf{s}_i) \wedge \mathbf{p}] = \mathbf{Pr}[\mathbf{f}_n = \mathbf{s}_j | \mathbf{f}_{n+1} = \mathbf{s}_i].$$

Since a Markov process observed in reverse order remains a Markov process, it might be suspected that the same is true for a Markov chain. This would be the case if the "backward transition probabilities," $p^*_{ij}(n) = \mathbf{Pr}[\mathbf{f}_n = \mathbf{s}_j | \mathbf{f}_{n+1} = \mathbf{s}_i]$, were independent of n. These probabilities may be found as follows:

$$\begin{aligned} p^*_{ij}(n) &= \frac{\mathbf{Pr}[\mathbf{f}_n = \mathbf{s}_j \wedge \mathbf{f}_{n+1} = \mathbf{s}_i]}{\mathbf{Pr}[\mathbf{f}_{n+1} = \mathbf{s}_i]} \\ &= \frac{\mathbf{Pr}[\mathbf{f}_{n+1} = \mathbf{s}_i | \mathbf{f}_n = \mathbf{s}_j] \cdot \mathbf{Pr}[\mathbf{f}_n = \mathbf{s}_j]}{\mathbf{Pr}[\mathbf{f}_{n+1} = \mathbf{s}_i]} \\ &= \frac{p_{ji} \cdot \mathbf{Pr}[\mathbf{f}_n = \mathbf{s}_j]}{\mathbf{Pr}[\mathbf{f}_{n+1} = \mathbf{s}_i]}. \end{aligned}$$

These transition probabilities would be independent of n only if the probability of being in a particular state at time n was independent of n. This is certainly not the case in general. For example, if the system is started in state s_1 with probability 1, then the probability that it is there on the next step is p_{11}. Thus, in general, $\mathbf{Pr}[\mathbf{f}_0 = \mathbf{s}_1] \neq \mathbf{Pr}[\mathbf{f}_1 = \mathbf{s}_1]$. Thus a Markov chain looked at in reverse order will be a Markov process, but in general its transition probabilities will depend on time and hence it will not be a Markov chain. We will return to this problem in § 5.3.

§2.2 Examples. In this section we shall give several simple examples of Markov chains which will be used in future work for illustrative purposes. The first five examples relate to what is normally called a "random walk." We imagine a particle which moves in a straight line in unit steps. Each step is one unit to the right with probability p or one unit to the left with probability q. It moves until it reaches one of two extreme points which are called "boundary points." The possibilities for its behavior at these points determine several different kinds of Markov chains. The states are the possible positions. We take the case of 5 states, states s_1 and s_5 being the "boundary" states, and s_2, s_3, s_4 the "interior states."

$$\begin{array}{ccccc} s_1 & s_2 & s_3 & s_4 & s_5 \\ \cdot & \cdot & \cdot & \cdot & \cdot \end{array}$$

EXAMPLE 1

Assume that if the process reaches state s_1 or s_5 it remains there from that time on. In this case the transition matrix is given by

$$P = \begin{array}{c} \\ s_1 \\ s_2 \\ s_3 \\ s_4 \\ s_5 \end{array} \begin{array}{c} \begin{array}{ccccc} s_1 & s_2 & s_3 & s_4 & s_5 \end{array} \\ \begin{pmatrix} 1 & 0 & 0 & 0 & 0 \\ q & 0 & p & 0 & 0 \\ 0 & q & 0 & p & 0 \\ 0 & 0 & q & 0 & p \\ 0 & 0 & 0 & 0 & 1 \end{pmatrix} \end{array}. \qquad (1)$$

EXAMPLE 2

Assume now that the particle is "reflected" when it reaches a boundary point and returns to the point from which it came. Thus if it ever hits s_1 it goes on the next step back to s_2. If it hits s_5 it goes on the next step back to s_4. The matrix of transition probabilities becomes in this case

$$P = \begin{array}{c} \\ s_1 \\ s_2 \\ s_3 \\ s_4 \\ s_5 \end{array} \begin{array}{c} \begin{array}{ccccc} s_1 & s_2 & s_3 & s_4 & s_5 \end{array} \\ \begin{pmatrix} 0 & 1 & 0 & 0 & 0 \\ q & 0 & p & 0 & 0 \\ 0 & q & 0 & p & 0 \\ 0 & 0 & q & 0 & p \\ 0 & 0 & 0 & 1 & 0 \end{pmatrix} \end{array}. \qquad (2)$$

EXAMPLE 3

As a third possibility we assume that whenever the particle hits one of the boundary states, it goes directly to the center state s_3. We may think of this as the process of Example 1 started at state s_3 and repeated each time the boundary is reached. The transition matrix is

$$P = \begin{array}{c} \\ s_1 \\ s_2 \\ s_3 \\ s_4 \\ s_5 \end{array} \begin{array}{c} \begin{array}{ccccc} s_1 & s_2 & s_3 & s_4 & s_5 \end{array} \\ \begin{pmatrix} 0 & 0 & 1 & 0 & 0 \\ q & 0 & p & 0 & 0 \\ 0 & q & 0 & p & 0 \\ 0 & 0 & q & 0 & p \\ 0 & 0 & 1 & 0 & 0 \end{pmatrix} \end{array}. \tag{3}$$

EXAMPLE 4

Assume now that once a boundary state is reached the particle stays at this state with probability $1/2$ and moves to the other boundary state with probability $1/2$. In this case the transition matrix is

$$P = \begin{array}{c} \\ s_1 \\ s_2 \\ s_3 \\ s_4 \\ s_5 \end{array} \begin{array}{c} \begin{array}{ccccc} s_1 & s_2 & s_3 & s_4 & s_5 \end{array} \\ \begin{pmatrix} 1/2 & 0 & 0 & 0 & 1/2 \\ q & 0 & p & 0 & 0 \\ 0 & q & 0 & p & 0 \\ 0 & 0 & q & 0 & p \\ 1/2 & 0 & 0 & 0 & 1/2 \end{pmatrix} \end{array}. \tag{4}$$

EXAMPLE 5

As the final choice for the behavior at the boundary, let us assume that when the particle reaches one boundary it moves directly to the other. The transition matrix is

$$P = \begin{array}{c} \\ s_1 \\ s_2 \\ s_3 \\ s_4 \\ s_5 \end{array} \begin{array}{c} \begin{array}{ccccc} s_1 & s_2 & s_3 & s_4 & s_5 \end{array} \\ \begin{pmatrix} 0 & 0 & 0 & 0 & 1 \\ q & 0 & p & 0 & 0 \\ 0 & q & 0 & p & 0 \\ 0 & 0 & q & 0 & p \\ 1 & 0 & 0 & 0 & 0 \end{pmatrix} \end{array}. \tag{5}$$

EXAMPLE 6

We next consider a modified version of the random walk. If the process is in one of the three interior states, it has equal probability of moving right, moving left, or staying in its present state. If it is on the boundary, it cannot stay, but has equal probability of moving to any of the four other states. The transition matrix is:

$$P = \begin{matrix} & \begin{matrix} \mathbf{s}_1 & \mathbf{s}_2 & \mathbf{s}_3 & \mathbf{s}_4 & \mathbf{s}_5 \end{matrix} \\ \begin{matrix} \mathbf{s}_1 \\ \mathbf{s}_2 \\ \mathbf{s}_3 \\ \mathbf{s}_4 \\ \mathbf{s}_5 \end{matrix} & \begin{pmatrix} 0 & 1/4 & 1/4 & 1/4 & 1/4 \\ 1/3 & 1/3 & 1/3 & 0 & 0 \\ 0 & 1/3 & 1/3 & 1/3 & 0 \\ 0 & 0 & 1/3 & 1/3 & 1/3 \\ 1/4 & 1/4 & 1/4 & 1/4 & 0 \end{pmatrix} \end{matrix}. \qquad (6)$$

EXAMPLE 7

A sequence of digits is generated at random. We take as states the following: $\mathbf{s}_1$ if a 0 occurs, $\mathbf{s}_2$ if a 1 or 2 occurs, $\mathbf{s}_3$ if a 3, 4, 5, or 6 occurs, $\mathbf{s}_4$ if a 7 or 8 occurs, $\mathbf{s}_5$ if a 9 occurs. This process is an independent trials process, but we shall see that Markov chain theory gives us information even about this special case. The transition matrix is

$$P = \begin{matrix} & \begin{matrix} \mathbf{s}_1 & \mathbf{s}_2 & \mathbf{s}_3 & \mathbf{s}_4 & \mathbf{s}_5 \end{matrix} \\ \begin{matrix} \mathbf{s}_1 \\ \mathbf{s}_2 \\ \mathbf{s}_3 \\ \mathbf{s}_4 \\ \mathbf{s}_5 \end{matrix} & \begin{pmatrix} .1 & .2 & .4 & .2 & .1 \\ .1 & .2 & .4 & .2 & .1 \\ .1 & .2 & .4 & .2 & .1 \\ .1 & .2 & .4 & .2 & .1 \\ .1 & .2 & .4 & .2 & .1 \end{pmatrix} \end{matrix}. \qquad (7)$$

EXAMPLE 8

According to *Finite Mathematics* (Chapter V, Section 8), in the Land of Oz they never have two nice days in a row. If they have a nice day they are just as likely to have snow as rain the next day. If they have snow (or rain) they have an even chance of having the same the next day. If there is a change from snow or rain, only half of the time is this a change to a nice day. We form a three-state Markov chain with states **R**, **N**, and **S** for rain, nice, and snow, respectively. The transition matrix is then

$$P = \begin{matrix} & \begin{matrix} \mathbf{R} & \mathbf{N} & \mathbf{S} \end{matrix} \\ \begin{matrix} \mathbf{R} \\ \mathbf{N} \\ \mathbf{S} \end{matrix} & \begin{pmatrix} 1/2 & 1/4 & 1/4 \\ 1/2 & 0 & 1/2 \\ 1/4 & 1/4 & 1/2 \end{pmatrix} \end{matrix}. \qquad (8)$$

In our previous examples the Markov property clearly held. In this case it could only be regarded as an approximation since the knowledge of the weather the last two days, for example, might lead us to different predictions than knowing the weather only on the previous day. One way to improve this approximation is to take as states the weather for two successive days. The states would then be **NN**, **NR**, **NS**, **RN**, **RR**, **RS**, **SN**, **SR**, **SS**. New transition probabilities would have to be estimated. A single step would still be one day, so that from **NR**, for example, we could move only to states **RN**, **RR**, **RS**. In examples of this kind it is possible to improve the approximation, still using the Markov chain theory, but at the expense of increasing the number of states.

EXAMPLE 9

An urn contains two unpainted balls. At a sequence of times a ball is chosen at random, painted either red or black, and put back. If the ball was unpainted, the choice of color is made at random. If it is painted, its color is changed. We form a Markov chain by taking as a state three numbers (x, y, z) where x is the number of unpainted balls, y the number of red balls, and z the number of black balls. The transition matrix is then

$$\begin{array}{c} \\ (0,1,1) \\ (0,2,0) \\ (0,0,2) \\ (2,0,0) \\ (1,1,0) \\ (1,0,1) \end{array} \begin{array}{c} \begin{array}{cccccc} (0,1,1) & (0,2,0) & (0,0,2) & (2,0,0) & (1,1,0) & (1,0,1) \end{array} \\ \begin{pmatrix} 0 & 1/2 & 1/2 & 0 & 0 & 0 \\ 1 & 0 & 0 & 0 & 0 & 0 \\ 1 & 0 & 0 & 0 & 0 & 0 \\ 0 & 0 & 0 & 0 & 1/2 & 1/2 \\ 1/4 & 1/4 & 0 & 0 & 0 & 1/2 \\ 1/4 & 0 & 1/4 & 0 & 1/2 & 0 \end{pmatrix} \end{array}. \qquad (9)$$

EXAMPLE 10

Assume that a student going to a certain college has each year a probability p of flunking out, a probability q of having to repeat the year, and a probability r of passing on to the next year. We form a Markov chain, taking as states s_1—has flunked out, s_2—has graduated, s_3—is a senior, s_4—is a junior, s_5—is a sophomore, s_6—is a freshman.

The transition matrix is then

$$P = \begin{array}{c} \\ s_1 \\ s_2 \\ s_3 \\ s_4 \\ s_5 \\ s_6 \end{array} \begin{array}{c} \begin{array}{cccccc} s_1 & s_2 & s_3 & s_4 & s_5 & s_6 \end{array} \\ \begin{pmatrix} 1 & 0 & 0 & 0 & 0 & 0 \\ 0 & 1 & 0 & 0 & 0 & 0 \\ p & r & q & 0 & 0 & 0 \\ p & 0 & r & q & 0 & 0 \\ p & 0 & 0 & r & q & 0 \\ p & 0 & 0 & 0 & r & q \end{pmatrix} \end{array}. \tag{10}$$

EXAMPLE 11

A man is playing two slot-machines. The first machine pays off with probability c, the second with probability d. If he loses, he plays the same machine again; if he wins, he switches to the other machine. Let s_i be the state of playing the i-th machine. The transition matrix is

$$P = \begin{array}{c} \\ s_1 \\ s_2 \end{array} \begin{array}{c} \begin{array}{cc} s_1 & s_2 \end{array} \\ \begin{pmatrix} 1-c & c \\ d & 1-d \end{pmatrix} \end{array}. \tag{11}$$

As c and d take on all permissible values ($0 \leqslant c \leqslant 1$, $0 \leqslant d \leqslant 1$) we get all 2×2 Markov chains.

EXAMPLE 12

Consider the special two-state Markov chain (Example 11) with transition matrix

$$P = \begin{array}{c} \\ s_1 \\ s_2 \end{array} \begin{array}{c} \begin{array}{cc} s_1 & s_2 \end{array} \\ \begin{pmatrix} 1/2 & 1/2 \\ 1/4 & 3/4 \end{pmatrix} \end{array}. \tag{11a}$$

(This can be called Example 11a.)

From this Markov chain we form a new Markov chain as follows. A state in the new chain will be a pair of states in the old chain. That is, the states are s_1s_1, s_1s_2, s_2s_1, s_2s_2. The new chain is in state s_is_j on the n-th step if the old chain was in state s_i on the n-th step and s_j on the $(n+1)$-th step.

The transition matrix for the new chain (Example 12) is

$$P = \begin{array}{c} \\ s_1s_1 \\ s_1s_2 \\ s_2s_1 \\ s_2s_2 \end{array} \begin{array}{c} \begin{array}{cccc} s_1s_1 & s_1s_2 & s_2s_1 & s_2s_2 \end{array} \\ \begin{pmatrix} 1/2 & 1/2 & 0 & 0 \\ 0 & 0 & 1/4 & 3/4 \\ 1/2 & 1/2 & 0 & 0 \\ 0 & 0 & 1/4 & 3/4 \end{pmatrix} \end{array}. \tag{12}$$

We shall see in § 6.5 that the study of this new chain gives us more detailed information about the original process than could be obtained directly from the two-state chain.

§ 2.3 Connection with matrix theory. In this section we shall show the connection between Markov chain theory and matrix theory. We shall start with the general finite Markov process and then specialize our results to the finite Markov chain.

2.3.1 THEOREM. *Let* $\mathbf{f}_n$ *be the outcome function at time n for a finite Markov process with transition probabilities* $p_{ij}(n)$, *then*

$$\mathbf{Pr}[\mathbf{f}_n = \mathbf{s}_v] = \sum_u \mathbf{Pr}[\mathbf{f}_{n-1} = \mathbf{s}_u] p_{uv}(n).$$

PROOF. The statement $\mathbf{f}_n = \mathbf{s}_v$ is a statement relative to the tree $\mathbf{T}_n$. To find its probability, we add the weights of all paths in its truth set. That is, all possible paths which end in outcome $\mathbf{s}_v$. Thus if $j, k, \ldots, u$ is a possible sequence of states

$$\mathbf{Pr}[\mathbf{f}_n = \mathbf{s}_v]$$

$$= \sum_{j,k,\ldots,u} \mathbf{Pr}[\mathbf{f}_0 = \mathbf{s}_j \wedge \cdots \wedge \mathbf{f}_{n-1} = \mathbf{s}_u \wedge \mathbf{f}_n = \mathbf{s}_v].$$

$$= \sum_{j,k,\ldots,u} \mathbf{Pr}[\mathbf{f}_0 = \mathbf{s}_j \wedge \cdots \wedge \mathbf{f}_{n-1} = \mathbf{s}_u] \cdot \mathbf{Pr}[\mathbf{f}_n = \mathbf{s}_v | \mathbf{f}_0 = \mathbf{s}_j \wedge \cdots \wedge \mathbf{f}_{n-1} = \mathbf{s}_u].$$

By the Markov property this is

$$\sum_{j,k,\ldots,u} \mathbf{Pr}[\mathbf{f}_0 = \mathbf{s}_j \wedge \cdots \wedge \mathbf{f}_{n-1} = \mathbf{s}_u] p_{uv}(n).$$

If in this last sum we keep u fixed and sum over the remaining indices we obtain

$$\mathbf{Pr}[\mathbf{f}_n = \mathbf{s}_v] = \sum_u \mathbf{Pr}[\mathbf{f}_{n-1} = \mathbf{s}_u] p_{uv}(n).$$

This completes the proof.

We can write the result of this theorem in matrix form. Let π_n

be a row vector which gives the induced measure for the outcome function $\mathbf{f}_n$. That is

$$\pi_n = \{p^{(n)}{}_1, p^{(n)}{}_2, \ldots, p^{(n)}{}_r\},$$

where $p^{(n)}{}_j = \mathbf{Pr}[\mathbf{f}_n = \mathbf{s}_j]$. Thus, $p^{(n)}{}_j$ is the probability that the process will after n steps be in state $\mathbf{s}_j$. The vector π_0 is the initial probabilities vector. Let $P(n)$ be the matrix with entries $p_{ij}(n)$. Then the result of Theorem 2.3.1 may be written in the form

$$\pi_n = \pi_{n-1} \cdot P(n)$$

for $n \geqslant 1$. By successive application of this result we have

$$\pi_n = \pi_0 \cdot P(1) \cdot P(2) \cdot \ldots \cdot P(n).$$

In the case of a Markov chain process, all the $P(n)$'s are the same and we obtain the following fundamental theorem.

2.3.2 THEOREM. *Let π_n be the induced measure for the outcome function $\mathbf{f}_n$ for a finite Markov chain with initial probability vector π_0 and transition matrix P. Then*

$$\pi_n = \pi_0 \cdot P^n.$$

This theorem shows that the key to the study of the induced measures for the outcome functions of a finite Markov chain is the study of the powers of the transition matrix. The entries of these powers have themselves an interesting probabilistic interpretation. To see this,

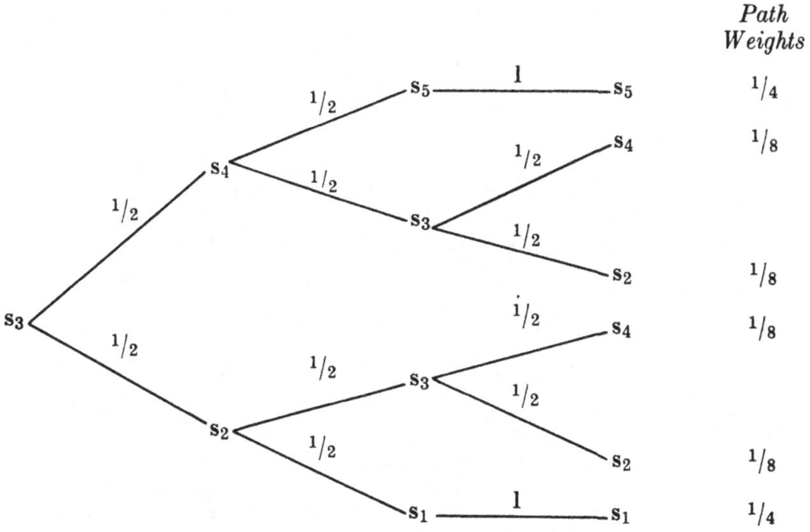

FIGURE 2-1

take as initial vector π_0 the vector with 1 in the i-th component and 0 otherwise. Then by Theorem 3.2, $\pi_n = \pi_0 P^n$. But $\pi_0 P^n$ is the i-th row of the matrix P^n. Thus the i-th row of the n-th power of the transition matrix gives the probability of being in each of the various states under the assumption that the process started in state $\mathbf{s}_i$.

In Example 1, let us assume that the process starts in state $\mathbf{s}_3$. Then $\pi_0 = \{0, 0, 1, 0, 0\}$. We can find the induced measures (see § 1.7) for the first three outcome functions by constructing a tree and tree measure for the first three experiments. This tree is given in Figure 2-1.

From this tree and tree measure we easily compute the induced measures for the functions $\mathbf{f}_1$, $\mathbf{f}_2$, $\mathbf{f}_3$. They are

$$\begin{aligned}\pi_1 &= \{0, {}^1/_2, 0, {}^1/_2, 0\}\\ \pi_2 &= \{{}^1/_4, 0, {}^1/_2, 0, {}^1/_4\}\\ \pi_3 &= \{{}^1/_4, {}^1/_4, 0, {}^1/_4, {}^1/_4\}.\end{aligned}$$

By Theorem 2.3.2 these induced measures should also be the third row in the matrices P, P^2, and P^3, since the starting state was $\mathbf{s}_3$. These matrices are

$$P = \begin{pmatrix} 1 & 0 & 0 & 0 & 0\\ {}^1/_2 & 0 & {}^1/_2 & 0 & 0\\ 0 & {}^1/_2 & 0 & {}^1/_2 & 0\\ 0 & 0 & {}^1/_2 & 0 & {}^1/_2\\ 0 & 0 & 0 & 0 & 1 \end{pmatrix}$$

$$P^2 = \begin{pmatrix} 1 & 0 & 0 & 0 & 0\\ {}^1/_2 & {}^1/_4 & 0 & {}^1/_4 & 0\\ {}^1/_4 & 0 & {}^1/_2 & 0 & {}^1/_4\\ 0 & {}^1/_4 & 0 & {}^1/_4 & {}^1/_2\\ 0 & 0 & 0 & 0 & 1 \end{pmatrix}$$

$$P^3 = \begin{pmatrix} 1 & 0 & 0 & 0 & 0\\ {}^5/_8 & 0 & {}^1/_4 & 0 & {}^1/_8\\ {}^1/_4 & {}^1/_4 & 0 & {}^1/_4 & {}^1/_4\\ {}^1/_8 & 0 & {}^1/_4 & 0 & {}^5/_8\\ 0 & 0 & 0 & 0 & 1 \end{pmatrix}.$$

We thus see that these matrices furnish us several tree measures simultaneously.

§ **2.4 Classification of states and chains.** We wish to classify the states of a Markov chain according to whether it is possible to go from a given state to another given state. This problem is exactly like the one treated in § 1.4. If we interpret **iTj** to mean that the process can go from state s_i to state s_j (not necessarily in one step), then all the results of that section are applicable.

In particular, the states are divided into equivalence classes. Two states are in the same equivalence class if they "communicate," i.e. if one can go from either state to the other one. The resulting partial ordering shows us the possible directions in which the process can proceed.

The minimal elements of the partial ordering are of particular interest.

2.4.1 DEFINITION. *The minimal elements of the partial ordering of equivalence classes are called* ergodic sets. *The remaining elements are called* transient sets. *The elements of a transient set are called* transient states. *The elements of an ergodic set are called* ergodic (*or* non-transient) states.

Since every finite partial ordering must have at least one minimal element, there must be at least one ergodic set for every Markov chain. However, there need be no transient set. The latter will occur if the entire chain consists of a single ergodic set, or if there are several ergodic sets which do not communicate with others.

From the results of § 1.4 we see that if a process leaves a transient set it can never return to this set, while if it once enters an ergodic set, it can never leave it. In particular, if an ergodic set contains only one element, then we have a state which once entered cannot be left. Such a state is called *absorbing*. Since from such a state we cannot go to another state, the following theorem characterizes absorbing states.

2.4.2 THEOREM. *A state s_i is absorbing if and only if $p_{ii} = 1$.*

It is convenient to use our classification to arrive at a canonical form for the transition matrix. We renumber the states as follows: The elements of a given equivalence class will receive consecutive numbers. The minimal sets will come first, then sets that are one level above the minimal sets, then sets two levels above the minimal sets, etc. This will assure us that we can go from a given state to another in the same class, or to a state in an earlier class, but not to a state in a later class. If the equivalence classes arranged as here described are $\mathbf{u}_1, \mathbf{u}_2, \ldots, \mathbf{u}_k$, then our matrix will appear as follows (where k is taken as 5, for the sake of illustration):

$$\begin{array}{l} \mathbf{u}_1: \\ \mathbf{u}_2: \\ \mathbf{u}_3: \\ \mathbf{u}_4: \\ \mathbf{u}_5: \end{array} \left(\begin{array}{c|c|c|c|c} P_1 & & & & \\ R_2 & P_2 & & O & \\ \multicolumn{2}{c|}{R_3} & P_3 & & \\ \multicolumn{3}{c|}{R_4} & P_4 & \\ \multicolumn{4}{c|}{R_5} & P_5 \end{array} \right).$$

Here the P_i represent transition matrices within a given equivalence class. The region O consists entirely of 0's. The matrix R_i will be entirely 0 if P_i is an ergodic set, but will have non-zero elements otherwise.

In this form it is easy to see what happens as P is raised to powers. Each power will be a matrix of the same form; in P^n we still have zeros in the upper region, and we simply have $P^n{}_i$ in the diagonal regions. This shows that a given equivalence class can be studied in isolation, by treating the submatrix P_i. This will be considered in detail later.

We can also apply the subdivision of an equivalence class considered in the previous chapter. We saw there that each equivalence class can be partitioned into cyclic classes. If there is only one cyclic class, then we say that the equivalence class is *regular*, otherwise we say that it is *cyclic*.

If an equivalence class is regular, then after sufficient time has elapsed the process can be in any state of the class, no matter which of the equivalent states it started in (see § 1.4). This means that all sufficiently high powers of its P_i must be positive (i.e. have only positive entries). If the equivalence class is cyclic, then no power of P_i can be positive.

From this classification of states we can arrive at a classification of Markov chains. We have noted that there must be an ergodic set, but there need be no transient set. This will lead to our primary subdivision. Within this we can subdivide according to the number and type of ergodic sets.

I. *Chains Without Transient Sets*

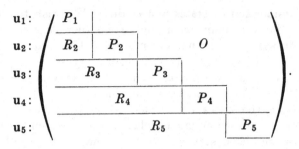

If such a chain has more than one ergodic set, then there is absolutely no interaction between these sets. Hence we have two or more unrelated Markov chains lumped together. These chains may be studied separately, and hence without loss of generality we may assume that the entire chain is a single ergodic set. A chain consisting of a single ergodic set is called an *ergodic chain.*

I-A. The ergodic set is regular. In this case the chain is called a *regular Markov chain.* As we see from previous considerations, all sufficiently high powers of P must be positive in this case. Thus no matter where the process starts, after sufficient lapse of time it could be in any state.

I-B. The ergodic set is cyclic. In this case the chain is called a *cyclic Markov chain.* Such a chain has a period d, and its states are subdivided into d cyclic sets ($d>1$). For a given starting position it will move through the cyclic sets in a definite order, returning to the set of the starting state after d steps. We also know that after sufficient time has elapsed, the process can be in any state of the cyclic set appropriate for the moment.

II. *Chains With Transient Sets*

In such a chain the process moves towards the ergodic sets. As will be seen in the next chapter, the probability that the process is in an ergodic set tends to 1; and it cannot escape from an ergodic set once it enters it. Hence it is fruitful to classify such chains by their ergodic sets.

II-A. All ergodic sets are unit sets. Such a chain is called an *absorbing chain.* In this case the process is eventually trapped in a single (absorbing) state. This type of process can also be characterized by the fact that all the ergodic states are absorbing states.

II-B. All ergodic sets are regular, but not all are unit sets.

II-C. All ergodic sets are cyclic.

II-D. There are both cyclic and regular ergodic sets.

Naturally, in each of these classes we can further classify chains according to how many ergodic sets there are. Of particular interest is the question whether there are one or more ergodic sets.

We can illustrate all of these types except II-D by the random walk examples.

For Example 1: The states s_1 and s_5 are absorbing states. The states s_2, s_3, s_4 are transient states. It is possible to go between any two of these states. Hence they form a single transient set. We have an absorbing Markov chain—that is, case II-A.

For Example 2: In this example it is possible to go from any state to any other state. Hence there are no transient states and there is a single ergodic set. Thus we have an ergodic chain. It is possible to return to a state only in an even number of steps. Thus the period of the states is 2. The two cyclic sets are $\{s_1, s_3, s_5\}$ and $\{s_2, s_4\}$. This is type I-B.

For Example 3: Again we can go from any state to any other state. Hence we again have an ergodic chain. It is possible to return to state s_3 from s_3 in either two or three steps. Hence the greatest common divisor $d=1$, and the period is 1. This is type I-A.

For Example 4: In this example $\{s_1, s_5\}$ is an ergodic set which is clearly regular. The set $\{s_2, s_3, s_4\}$ is the single transient set. This is type II-B.

For Example 5: Here we have a single ergodic set $\{s_1, s_5\}$ which has period 2. The set $\{s_2, s_3, s_4\}$ is again a transient set. This is type II-C.

§ **2.5 Problems to be studied.** Let us consider our various types of chains, and ask what types of problems we would like to answer in the following chapters.

First of all we may wish to study a regular Markov chain. In such a chain the process keeps moving through all the states, no matter where it starts. Some of the questions of interest are:

(1) If a chain starts in s_i, what is the probability after n steps that it will be in s_j?

(2) Can we predict the average number of times that the process is in s_i? And if so, how does this depend on where the process starts?

(3) We may wish to consider the process as it goes from s_i to s_j. What is the mean and variance of the number of steps needed? What are the mean and variance of the number of states passed? What is the probability that the process passes through s_k?

(4) We may wish to study a certain subset of states, and observe the process only when it is in these states. How does this modify our previous results? These questions are treated in Chapter IV.

Next we may wish to study a cyclic chain. Here the same kinds of questions are of interest as for a regular chain. Naturally, a regular chain is easier to study; and we will find that, once we have the answers for regular chains, it is not hard to find the corresponding answers for all ergodic chains. This extension of regular chain theory to the theory of ergodic chains is carried out in Chapter V.

Next we may wish to consider a Markov chain with transient states. There are two kinds of questions to be asked here. One will concern the behavior of the chain before it enters an ergodic set, while the other kind will apply after the chain has entered an ergodic set. The latter questions are no different from the ones considered above. Once a chain enters an ergodic set it can never leave it, and hence the existence of states outside the set is irrelevant. Thus questions of the second kind can be answered by considering a chain consisting of a single ergodic set, i.e. an ergodic chain.

The really new questions concern the behavior of the chain up to the moment that it enters an ergodic set. However, for these questions the nature of the ergodic states is irrelevant, and we may make them all into absorbing states if we wish. More generally, if we wish to study the process while it is in a set of transient states, we may make all other states absorbing. This modified process will serve to find all the answers we desire. Hence the only new questions concern the behavior of an absorbing chain.

Some of the questions that are of interest concerning a transient state s_i are:

(1) The probability of entering a given ergodic set, starting from s_i.

(2) The mean and variance of the number of times that the process is in s_i before entering an ergodic set, and how this number depends on the starting position.

(3) The mean and variance of the number of steps needed before entering an ergodic set starting at s_i.

(4) The mean number of states passed before entering an ergodic set, starting at s_i.

Chapter III will deal with absorbing chains, and all these questions will be answered. Thus we will find the most interesting questions about finite Markov chains answered in Chapters III, IV, and V.

Exercises for Chapter II

For § 2.1

1. Five points are marked on a circle. A process moves from a given point to one of its neighbors, with probability $1/2$ for each neighbor. Find the transition matrix of the resulting Markov chain.

2. Three tanks fight a duel. Tank **A** hits its target with probability $2/3$, tank **B** with probability $1/2$, and tank **C** with probability $1/3$. Shots are fired simultaneously, and once a tank is hit it is out of action. As a state we choose the set of tanks still in action. If on each step each tank fires at its strongest opponent, verify that the following transition matrix is correct:

	E	A	B	C	AC	BC	ABC
E	1	0	0	0	0	0	0
A	0	1	0	0	0	0	0
B	0	0	1	0	0	0	0
C	0	0	0	1	0	0	0
AC	$2/9$	$4/9$	0	$1/9$	$2/9$	0	0
BC	$1/6$	0	$1/3$	$1/6$	0	$1/3$	0
ABC	0	0	0	$4/9$	$2/9$	$2/9$	$1/9$

3. Modify the transition matrix in the previous exercise, assuming that when all tanks are in action, **A** fires at **B**, **B** at **C**, and **C** at **A**.

4. We carry out a sequence of experiments as follows: At first a fair coin is tossed. Then, if experiment $n-1$ comes out heads, we toss a fair coin; if it comes out tails, we toss a coin which has probability $1/n$ of coming up heads. What are the transition probabilities? What kind of process is this?

For § 2.2

5. Modify Example 1 by assuming that when the process reaches s_1 it goes on the next step to state s_2. Form the new transition matrix.

6. Modify the process described in Example 2 by assuming that when the process reaches s_1 it stays there for the next two steps and on the third step moves to state s_2. Show that the resulting process is not a Markov chain (with the five given states).

7. In Exercise 6 show that we can treat the process as a Markov chain, by allowing a larger number of states. Write down the transition matrix.

8. Modify the transition matrix of Example 7, assuming that the digit 0 is twice as likely to be generated than any other digit.

9. Modify Example 7, assuming that the same digit is never generated twice in a row, but otherwise digits are equally likely to occur.

10. In Example 8 allow only two states: Nice and not nice. Show that the process is still a Markov chain, and find its transition matrix.

For § 2.3

11. In Example 11a compute P^2, P^4, P^8, and P^{16}, and write the entries as decimal fractions. Note the trend, and interpret your results.

12. Show that, no matter how Example 7 is started, the probabilities for being in each of the states after 1 step agree with the common row for the transition matrix. What are the probabilities after n steps?

13. Assume that Example 8 is started with initial vector $\pi_0 = (^2/_5, ^1/_5, ^2/_5)$. Find π_1, π_2. What is π_n?

14. The weather is nice today in the Land of Oz. What kind of weather is most likely to occur day after tomorrow?

15. In Example 11, assume that $c = ^1/_2$ and $d = ^1/_4$. The man randomly chooses the machine to play first. What is the probability that he plays the better machine (a) on the second play, (b) on the third play, and (c) on the fourth play?

16. In Example 2 assume that the process is started in state s_3. Construct a tree and tree measure for the first three experiments. Use this to find the induced measure for the first three outcome functions. Verify that your results agree with the probabilities found from P, P^2, and P^3.

For § 2.4

17. For the following Markov chain, give a complete classification of the states and put the transition matrix in canonical form.

$$P = \begin{array}{c} \\ s_1 \\ s_2 \\ s_3 \\ s_4 \\ s_5 \\ s_6 \\ s_7 \end{array} \begin{array}{c} \begin{array}{ccccccc} s_1 & s_2 & s_3 & s_4 & s_5 & s_6 & s_7 \end{array} \\ \begin{pmatrix} 0 & 0 & 0 & 1 & 0 & 0 & 0 \\ 0 & 0 & 0 & 0 & 0 & 0 & 1 \\ 0 & 0 & 1 & 0 & 0 & 0 & 0 \\ 1/2 & 0 & 0 & 1/2 & 0 & 0 & 0 \\ 0 & 0 & 0 & 0 & 1 & 0 & 0 \\ 0 & 0 & 0 & 0 & 0 & 0 & 1 \\ 0 & 1/2 & 0 & 0 & 0 & 1/2 & 0 \end{pmatrix} \end{array}.$$

18. A Markov chain has the following transition matrix, with non-zero entries marked by x. Give a complete classification of the states and put the transition matrix in canonical form.

$$P = \begin{array}{c} \\ s_1 \\ s_2 \\ s_3 \\ s_4 \\ s_5 \\ s_6 \\ s_7 \\ s_8 \\ s_9 \end{array} \begin{array}{c} \begin{array}{ccccccccc} s_1 & s_2 & s_3 & s_4 & s_5 & s_6 & s_7 & s_8 & s_9 \end{array} \\ \begin{pmatrix} x & x & 0 & 0 & 0 & 0 & 0 & 0 & 0 \\ x & x & x & 0 & 0 & 0 & x & 0 & 0 \\ 0 & 0 & 0 & 0 & 0 & 0 & x & 0 & 0 \\ 0 & 0 & 0 & x & 0 & 0 & 0 & 0 & x \\ 0 & 0 & 0 & 0 & x & 0 & 0 & 0 & 0 \\ 0 & 0 & x & 0 & 0 & x & 0 & 0 & 0 \\ 0 & 0 & x & 0 & 0 & 0 & 0 & 0 & 0 \\ 0 & x & 0 & 0 & 0 & x & 0 & x & 0 \\ 0 & 0 & 0 & x & 0 & 0 & 0 & 0 & x \end{pmatrix} \end{array}.$$

19. Classify the following chains as ergodic or absorbing. Which of the ergodic chains is regular?

(a) $P = \begin{pmatrix} 1/2 & 1/2 \\ 1/2 & 1/2 \end{pmatrix}$

(b) $P = \begin{pmatrix} 1 & 0 & 0 \\ 0 & 1 & 0 \\ 1/3 & 1/3 & 1/3 \end{pmatrix}$

(c) $P = \begin{pmatrix} 0 & 1/2 & 1/2 \\ 1 & 0 & 0 \\ 1 & 0 & 0 \end{pmatrix}$

(d) $P = \begin{pmatrix} 1 & 0 & 0 & 0 \\ 0 & 1/3 & 1/3 & 1/3 \\ 0 & 1/3 & 1/3 & 1/3 \\ 0 & 1/3 & 1/3 & 1/3 \end{pmatrix}$

(e) $P = \begin{pmatrix} 0 & 1 & 0 \\ 0 & 0 & 1 \\ 1/2 & 1/2 & 0 \end{pmatrix}$

20. In Example 9 classify the states. Put the transition matrix in canonical form. What type of chain is this?

21. For an ergodic chain the i-th state is made absorbing by replacing the i-th row in the transition matrix by a row with a 1 in the i-th component. Prove that the resulting chain is absorbing.

22. In Example 11, give conditions on c and d so that the resulting chain is

(a) ergodic (b) regular (c) cyclic (d) absorbing

For the entire chapter

23. In a certain state a voter is allowed to change his party affiliation (for primary elections) only by abstaining from the primary for one year. Let s_1 indicate that a man votes Democratic, s_2 that he votes Republican, and s_3 that he abstains, in the given year. Experience shows that a Democrat will abstain $1/2$ the time in the following primary, a Republican will abstain $1/4$ time, while a voter who abstained for a year is equally likely to vote for either party in the next election. [We will refer to this as Example 13.]

(a) Find the transition matrix.
(b) Find the probability that a man who votes Democratic this year will abstain three years from now.
(c) Classify the states.
(d) In a given year $1/4$ of the population votes Democratic, $1/2$ Republican, the rest abstain. What proportions do you expect in the next primary election?

24. A sequence of experiments is performed, in each of which two fair coins are tossed. Let s_1 indicate that two heads come up, s_2 that a head and a tail come up, and s_3 that two tails turn up. [We will refer to this as Example 14.]

(a) Find the transition matrix.
(b) If two heads turn up on a given toss, what is the probability of two heads turning up three tosses later?
(c) Classify the states.

CHAPTER III

ABSORBING MARKOV CHAINS

§ **3.1 Introduction.** Let us recall the basic definitions relevant to an absorbing chain. In the classification of states, the equivalence classes were divided into transient and ergodic sets. The former, once left, are never again entered; while the latter, once entered, are never again left. If a state is the only element of an ergodic set, then it is called an absorbing state. For such a state s_i the entry p_{ii} must be 1, and hence all other entries in this row of the transition matrix are 0. A chain, all of whose non-transient states are absorbing, is called an absorbing chain. These chains will occupy us in the present chapter.

3.1.1 THEOREM. *In any finite Markov chain, no matter where the process starts, the probability after n steps that the process is in an ergodic state tends to* 1 *as n tends to infinity.*

PROOF. If the process once reaches an ergodic state, then it can never leave its equivalence class, and hence it will at all future steps be in an ergodic state. Suppose that it starts in a transient state. Its equivalence class is not minimal; hence there is a minimal element below it. This means that it must be possible to reach some ergodic set. Let us suppose that from any transient state it is possible to reach an ergodic state in not more than n steps. (Since there are only a finite number of states, n is simply the maximum of the number of steps required from each state.) Hence there is a positive number p such that the probability of entering an ergodic state in at most n steps is at least p, from any transient state. Hence the probability of *not* reaching an ergodic state in n steps is at most $(1-p)$, which is less than 1. The probability of not reaching an ergodic state in kn steps is less than or equal to $(1-p)^k$, and this probability tends to 0 as k increases. Hence the theorem follows.

3.1.2 COROLLARY. *There are numbers* $b>0$, $0<c<1$ *such that* $p^{(n)}{}_{ij} \leqslant b \cdot c^n$, *for any transient states* s_i, s_j.

This is a direct consequence of the above proof. It shows the rate at which $p^{(n)}{}_{ij}$ tends to 0.

It is convenient to consider the canonical form of the matrix P in an aggregated version. We unite all the ergodic sets, and all the transient sets. (Let us say that there are s transient states, and $r-s$ ergodic states.) The form then becomes

$$P = \begin{pmatrix} \overbrace{S}^{r-s} & \overbrace{O}^{s} \\ R & Q \end{pmatrix} \begin{matrix} \}r-s \\ \}s \end{matrix}.$$

Here again the region O consists entirely of 0's. The $s \times s$ submatrix Q concerns the process as long as it stays in transient states, the $s \times (r-s)$ matrix R concerns the transition from transient to ergodic states, and the $(r-s) \times (r-s)$ matrix deals with the process after it has reached an ergodic set. From Theorem 3.1.1 we see that the powers of Q tend to O. Hence as we raise P to higher and higher powers, the matrices approach a matrix whose last s columns are all 0. This is the matrix version of Theorem 3.1.1.

Let us now consider an absorbing chain. By its definition we see that S is $I_{(r-s)\times(r-s)}$, i.e. an identity matrix of the appropriate dimension. Thus its canonical form is

$$P = \begin{pmatrix} \overset{r-s}{I} & \overset{s}{O} \\ R & Q \end{pmatrix} \begin{matrix} r-s \\ s \end{matrix}.$$

And by the nature of the powers of P we know that the region I remains I. This corresponds to the fact that once an absorbing state is entered, it cannot be left. From Theorem 3.1.1 we know that the probability that such a state is entered, in an absorbing chain, tends to 1. Hence we may say that with probability 1 the chain will enter an absorbing state and stay there, i.e. that it will be "absorbed."

Let us write some of our examples from Chapter II, § 2.2, in the new canonical form. In Example 1 the states s_1 and s_5 are absorbing, hence these must be written first. We thus have

$$P = \begin{array}{c} \\ s_1 \\ s_5 \\ s_2 \\ s_3 \\ s_4 \end{array} \begin{array}{cc|ccc} s_1 & s_5 & s_2 & s_3 & s_4 \\ 1 & 0 & 0 & 0 & 0 \\ 0 & 1 & 0 & 0 & 0 \\ \hline q & 0 & 0 & p & 0 \\ 0 & 0 & q & 0 & p \\ 0 & p & 0 & q & 0 \end{array}$$

where the regions I, O, R, and Q have been marked off.

The matrix for Example 10 is already in canonical form in § 2.2. The first two states are absorbing. Hence R is 4×2 and Q is 4×4 in this example.

Example 9 is not an absorbing chain. It has a single ergodic set, consisting of the first three states. The matrix appears in canonical form in § 2.2. If we want to study this process only until it enters the ergodic set, then we may make the ergodic states absorbing. The resulting transition matrix is

$$P = \left(\begin{array}{ccc|ccc} 1 & 0 & 0 & 0 & 0 & 0 \\ 0 & 1 & 0 & 0 & 0 & 0 \\ 0 & 0 & 1 & 0 & 0 & 0 \\ \hline 0 & 0 & 0 & 0 & 1/2 & 1/2 \\ 1/4 & 1/4 & 0 & 0 & 0 & 1/2 \\ 1/4 & 0 & 1/4 & 0 & 1/2 & 0 \end{array}\right).$$

If we do not even care at which state the ergodic set is entered, we may lump the three ergodic states into a single one, obtaining the much simpler matrix

$$P = \left(\begin{array}{c|ccc} 1 & 0 & 0 & 0 \\ \hline 0 & 0 & 1/2 & 1/2 \\ 1/2 & 0 & 0 & 1/2 \\ 1/2 & 0 & 1/2 & 0 \end{array}\right).$$

The former matrix preserves Q and R, while it modifies S; the latter preserves only Q. This is in good agreement with the interpretation given for Q, R, and S earlier in this section.

It should now be clear that absorbing chains serve to answer all questions of the second type (concerning transient states) raised in § 2.5. But absorbing chains are also important in the study of an ergodic set. Suppose that we wish to ask a question about what happens as the process goes from s_i to s_j Then we may wish to "stop" the process as soon as it reaches s_j, which we accomplish by making s_j an absorbing state. And since s_j can be reached from all states of its equivalence class, the resulting chain will be an absorbing Markov chain. This trick will be developed in § 6.1.

§ **3.2 The fundamental matrix.** The following basic theorem is a direct consequence of the matrix theorem we proved in § 1.11.1, if we recall that Q^k tends to 0.

3.2.1 THEOREM. *For any absorbing Markov chain, $I-Q$ has an inverse, and*

$$(I-Q)^{-1} = I+Q+Q^2+\cdots = \sum_{k=0}^{\infty} Q^k.$$

3.2.2 DEFINITION. *For an absorbing Markov chain we define the* fundamental matrix *to be $N=(I-Q)^{-1}$.*

3.2.3 DEFINITION. *We define $\mathbf{n}_j$ to be the function giving the total number of times that the process is in $\mathbf{s}_j$. (This is defined only for transient state $\mathbf{s}_j$.) $\mathbf{u}^k{}_j$ is defined as the function that is 1 if the process is in state $\mathbf{s}_j$ after k steps, and is 0 otherwise. (See §§ 1.7 and 1.8 for the notation used in this section.)*

We will now give a probabilistic interpretation to N. We let $\mathbf{T}$ be the set of transient states.

3.2.4 THEOREM. $\{\mathbf{M}_i[\mathbf{n}_j]\}=N$, *where* $\mathbf{s}_i,\ \mathbf{s}_j \in \mathbf{T}$.

PROOF. It is easily seen that $\mathbf{n}_j = \sum_{k=0}^{\infty} \mathbf{u}^k{}_j$.

Hence

$$\begin{aligned}
\{\mathbf{M}_i[\mathbf{n}_j]\} &= \left\{\mathbf{M}_i\left[\sum_{k=0}^{\infty} \mathbf{u}^k{}_j\right]\right\} \\
&= \left\{\sum_{k=0}^{\infty} \mathbf{M}_i[\mathbf{u}^k{}_j]\right\} \\
&= \left\{\sum_{k=0}^{\infty} ((1-p^{(k)}{}_{ij})\cdot 0 + p^{(k)}{}_{ij}\cdot 1)\right\} \\
&= \sum_{k=0}^{\infty} \{p^{(k)}{}_{ij}\} \\
&= \sum_{k=0}^{\infty} Q^k \quad \text{since } \mathbf{s}_i,\ \mathbf{s}_j \text{ are transient} \\
&= N \quad \text{by 3.2.1, 3.2.2.}
\end{aligned}$$

This completes the proof.

This theorem establishes the fact that the mean of the total number of times the process is in a given transient state is always finite, and that these means are simply given by N.

There is an interesting alternative proof for this result. To compute $\mathbf{M}_i[\mathbf{n}_j]$, we may add up the original position's contribution, plus each of the steps' contribution. The original position contributes 1 if and only if $i=j$. It is convenient to define $\mathbf{d}_{ij}$, the constant function that is 1 if $i=j$, 0 otherwise. Then we can say that the original position contributes $\mathbf{d}_{ij}$. After one step we move to s_k with probability p_{ik}. If the new state is absorbing, it contributes nothing to our mean, but if it is transient, then it contributes $\mathbf{M}_k[\mathbf{n}_j]$. Hence we have

$$\mathbf{M}_i[\mathbf{n}_j] = \mathbf{d}_{ij} + \sum_{s_k \in \mathbf{T}} p_{ik}\mathbf{M}_k[\mathbf{n}_j]$$

$$\{\mathbf{M}_i[\mathbf{n}_j]\} = I + Q \cdot \{\mathbf{M}_i[\mathbf{n}_j]\}.$$

Hence

$$\{\mathbf{M}_i[\mathbf{n}_j]\} = (I-Q)^{-1} = N.$$

We will apply these results to the examples of the last section. In the random walk, Example 1 of § 2.2,

$$(I-Q) = \begin{pmatrix} 1 & -p & 0 \\ -q & 1 & -p \\ 0 & -q & 1 \end{pmatrix}$$

and hence

$$N = (I-Q)^{-1} = \begin{array}{c} \\ s_2 \\ s_3 \\ s_4 \end{array} \begin{array}{c} \begin{array}{ccc} s_2 & s_3 & s_4 \end{array} \\ \begin{pmatrix} \frac{p+q^2}{p^2+q^2} & \frac{p}{p^2+q^2} & \frac{p^2}{p^2+q^2} \\ \frac{q}{p^2+q^2} & \frac{1}{p^2+q^2} & \frac{p}{p^2+q^2} \\ \frac{q^2}{p^2+q^2} & \frac{q}{p^2+q^2} & \frac{q+p^2}{p^2+q^2} \end{pmatrix} \end{array}.$$

[Since $p+q=1$, and hence $(p+q)^2=1$, we have that $1-2pq=p^2+q^2$.] We see that, for example, if the process starts in s_3 (the middle state), then it will be in the middle state an average of $1/(p^2+q^2)$ times. This quantity is always between 1 and 2. The minimum of 1 is achieved if $p=0$ or 1, the maximum of 2 if $p={}^1/_2$. In the former case the process starts at s_3 and goes directly to one of the boundaries, hence it will be in state s_3 only at the beginning. But even in the case $p={}^1/_2$ we expect the process to return only once on the average.

3.2.5 EXAMPLE 1a. As an illustration we give the fundamental matrix for the case $p={}^2/_3$, i.e. when it is twice as likely to move to the right as to the left.

$$N = \begin{array}{c} \\ s_2 \\ s_3 \\ s_4 \end{array} \begin{array}{c} \begin{array}{ccc} s_2 & s_3 & s_4 \end{array} \\ \begin{pmatrix} {}^7/_5 & {}^6/_5 & {}^4/_5 \\ {}^3/_5 & {}^9/_5 & {}^6/_5 \\ {}^1/_5 & {}^3/_5 & {}^7/_5 \end{pmatrix} \end{array}.$$

In the college example, Example 10 of § 2.2, remembering that $p+q+r=1$, we have

$$(I-Q) = \begin{pmatrix} p+r & 0 & 0 & 0 \\ -r & p+r & 0 & 0 \\ 0 & -r & p+r & 0 \\ 0 & 0 & -r & p+r \end{pmatrix}.$$

$$N = (I-Q)^{-1} = \begin{array}{c} \\ s_3 \\ s_4 \\ s_5 \\ s_6 \end{array} \begin{array}{c} \begin{array}{cccc} s_3 & s_4 & s_5 & s_6 \end{array} \\ \begin{pmatrix} \frac{1}{p+r} & 0 & 0 & 0 \\ \frac{r}{(p+r)^2} & \frac{1}{p+r} & 0 & 0 \\ \frac{r^2}{(p+r)^3} & \frac{r}{(p+r)^2} & \frac{1}{p+r} & 0 \\ \frac{r^3}{(p+r)^4} & \frac{r^2}{(p+r)^3} & \frac{r}{(p+r)^2} & \frac{1}{p+r} \end{pmatrix} \end{array} . \begin{array}{l} \text{SENIOR} \\ \text{JUNIOR} \\ \text{SOPHOMORE} \\ \text{FRESHMAN} \end{array}$$

The zeros in N indicate that no one is demoted in the college. Thus, for example, a junior cannot spend any time as a sophomore or freshman in the future. As an illustration we compute N (approximately) for the case we will call Example 10a, where the probabilities of flunking out, repeating, and being promoted are $p=.2$, $q=.1$, $r=.7$, respectively.

$$N = \begin{pmatrix} 1.11 & 0 & 0 & 0 \\ .86 & 1.11 & 0 & 0 \\ .67 & .86 & 1.11 & 0 \\ .52 & .67 & .86 & 1.11 \end{pmatrix} \begin{array}{l} \text{SENIOR} \\ \text{JUNIOR} \\ \text{SOPHOMORE} \\ \text{FRESHMAN} \end{array}$$

In the urn example, Example 9 of § 2.2, we have

$$(I-Q) = \begin{pmatrix} 1 & -1/2 & -1/2 \\ 0 & 1 & -1/2 \\ 0 & -1/2 & 1 \end{pmatrix}$$

$$N = (I-Q)^{-1} = \begin{array}{c} s_4 \\ s_5 \\ s_6 \end{array} \begin{pmatrix} 1 & 1 & 1 \\ 0 & 4/3 & 2/3 \\ 0 & 2/3 & 4/3 \end{pmatrix} \begin{array}{l} (2,0,0) \\ (1,1,0). \\ (1,0,1) \end{array}$$

If the process reaches (1,1,0) or (1,0,1), then from then on it is expected to be in that state $^4/_3$ times, and in the other state $^2/_3$ times. (The $^4/_3$ includes the original position.) From neither of these states can the state (2,0,0) be reached, since a painted ball always remains painted. If the process starts in (2,0,0), which is its natural starting position, it will be in this position only once. It is expected to be in each of the other two states once, which is the average of $^4/_3$ and $^2/_3$.

These fundamental matrices will be used throughout this chapter for illustrations.

§ **3.3 Applications of the fundamental matrix.** We will show that a number of interesting quantities can be expressed in terms of the fundamental matrix. These results will here be illustrated in terms of the random walk Example 1a (see § 3.2.5), and all the absorbing chains will be worked out in the next section.

3.3.1 DEFINITION. *We define the following new matrices and vectors:*

$N_2 = N(2N_{dg}-I)-N_{sq}$	$s \times s$ *matrix*
$B = NR$	$s \times (r-s)$ *matrix*
$\tau = N\xi$	s *component column vector*
$\tau_2 = (2N-I)\tau-\tau_{sq}$	s *component column vector*

3.3.2 THEOREM. $QN = NQ = N-I$.

PROOF. From 3.2.1, 3.2.2,

$$N = I+Q+Q^2+\cdots.$$

Hence,

$$QN = NQ = Q+Q^2+Q^3+\cdots,$$

which is the original series without I.

3.3.3 THEOREM. $\{\mathbf{Var}_i[\mathbf{n}_j]\} = N_2$, *where* $\mathbf{s}_i$, $\mathbf{s}_j \in \mathbf{T}$.

PROOF. We recall that $\mathbf{Var}_i[\mathbf{n}_j] = \mathbf{M}_i[\mathbf{n}^2_j] - \mathbf{M}_i[\mathbf{n}_j]^2$. From Theorem 3.2.4 we see that

$$\{\mathbf{M}_i[\mathbf{n}_j]^2\} = N_{sq},$$

hence we need only show that

$$\{\mathbf{M}_i[\mathbf{n}^2_j]\} = N(2N_{dg}-I).$$

We will assume that these means are finite. A proof of this fact will be given at the end of this section. To compute these means we again ask where the process can go in one step, from its starting position $\mathbf{s}_i$. It can go to $\mathbf{s}_k$ with probability p_{ik}. If the new state is absorbing, then we can never reach $\mathbf{s}_j$ again, and the only possible contribution is from the initial state, which is $\mathbf{d}_{ij}$. If the new state is transient, we will be in $\mathbf{s}_j$ $\mathbf{d}_{ij}$ times from the original position, and $\mathbf{n}_j$ times from the later steps. Hence, remembering that $\mathbf{d}_{ij}$ is a constant function and $\mathbf{d}_{ij} = \mathbf{d}^2_{ij}$,

$$\begin{aligned}\{\mathbf{M}_i[\mathbf{n}^2_j]\} &= \left\{\sum_{s_k \in \tilde{T}} p_{ik}\mathbf{d}^2_{ij} + \sum_{s_k \in T} p_{ik}\mathbf{M}_k[(\mathbf{n}_j + \mathbf{d}_{ij})^2]\right\} \\ &= \left\{\sum_{s_k \in T} p_{ik}(\mathbf{M}_k[\mathbf{n}^2_j] + 2\mathbf{M}_k[\mathbf{n}_j]\cdot\mathbf{d}_{ij}) + \mathbf{d}_{ij}\right\} \\ &= Q\{\mathbf{M}_i[\mathbf{n}^2_j]\} + 2(QN)_{dg} + I.\end{aligned}$$

Hence

$$\begin{aligned}\{\mathbf{M}_i[\mathbf{n}^2_j]\} &= (I-Q)^{-1}(2(QN)_{dg} + I) \\ &= N(2(N-I)_{dg} + I) = N(2N_{dg} - I).\end{aligned}$$

The matrix $(QN)_{dg}$ appeared above since the factor $\mathbf{d}_{ij}$ has the effect of setting all elements off the main diagonal equal to 0.

In our Example 1a, we have already computed N, and we now compute N_2 as well.

$$N = \begin{pmatrix} 7/5 & 6/5 & 4/5 \\ 3/5 & 9/5 & 6/5 \\ 1/5 & 3/5 & 7/5 \end{pmatrix} \qquad N_{sq} = \begin{pmatrix} 49/25 & 36/25 & 16/25 \\ 9/25 & 81/25 & 36/25 \\ 1/25 & 9/25 & 49/25 \end{pmatrix}$$

$$N_{dg} = \begin{pmatrix} 7/5 & 0 & 0 \\ 0 & 9/5 & 0 \\ 0 & 0 & 7/5 \end{pmatrix} \qquad 2N_{dg} - I = \begin{pmatrix} 9/5 & 0 & 0 \\ 0 & 13/5 & 0 \\ 0 & 0 & 9/5 \end{pmatrix}$$

$$N(2N_{dg} - I) = \begin{pmatrix} 63/25 & 78/25 & 36/25 \\ 27/25 & 117/25 & 54/25 \\ 9/25 & 39/25 & 63/25 \end{pmatrix} \quad N_2 = \begin{matrix} \\ s_2 \\ s_3 \\ s_4 \end{matrix}\!\!\begin{matrix} \begin{matrix} s_2 & \;\; s_3 & \;\; s_4 \end{matrix} \\ \begin{pmatrix} 14/25 & 42/25 & 20/25 \\ 18/25 & 36/25 & 18/25 \\ 8/25 & 30/25 & 14/25 \end{pmatrix} \end{matrix}.$$

Thus we see that for any state as initial state the variance is largest for the middle state. We also note that N_2 is quite large compared to N_{sq}; hence the means are fairly unreliable estimates for this Markov chain. This will often be the case.

3.3.4 DEFINITION. *Let* $\mathbf{t}$ *be the function giving the number of steps (including the original position) in which the process is in a transient state.*

If the process starts in an ergodic state, then $\mathbf{t} = 0$. If the process starts in a transient state, then $\mathbf{t}$ gives the total number of steps needed to reach an ergodic set. In an absorbing chain this is the time to absorption.

3.3.5 THEOREM. $\{\mathbf{M}_i[\mathbf{t}]\} = \tau$; $\{\mathbf{Var}_i[\mathbf{t}]\} = \tau_2$, *where* $\mathbf{s}_i \in \mathbf{T}$.

PROOF. It is easily seen that $\mathbf{t} = \sum_{s_j \in \mathbf{T}} \mathbf{n}_j$.

Hence

$$\begin{aligned}\{\mathbf{M}_i[\mathbf{t}]\} &= \left\{\sum_{s_j \in \mathbf{T}} \mathbf{M}_i[\mathbf{n}_j]\right\} \\ &= N\xi,\end{aligned}$$

since this gives the row sums of N.

For the variance we carry out an argument similar to that in §3.3.3, but here the first step always counts.

$$\begin{aligned}\{\mathbf{M}_i[\mathbf{t}^2]\} &= \left\{\sum_{\mathbf{s}_k \in \tilde{\mathbf{T}}} p_{ik} \cdot 1 + \sum_{\mathbf{s}_k \in \mathbf{T}} p_{ik}\mathbf{M}_k[(\mathbf{t}+1)^2]\right\} \\ &= \left\{\sum_{\mathbf{s}_k \in \mathbf{T}} p_{ik}(\mathbf{M}_k[\mathbf{t}^2] + 2\mathbf{M}_k[\mathbf{t}]) + 1\right\} \\ &= Q\{\mathbf{M}_i[\mathbf{t}^2]\} + 2Q\tau + \xi.\end{aligned}$$

Hence

$$\begin{aligned}\{\mathbf{M}_i[\mathbf{t}^2]\} &= (I-Q)^{-1}(2Q\tau + \xi) \\ &= 2NQ\tau + N\xi \\ &= 2(N-I)\tau + \tau \\ &= (2N-I)\tau.\end{aligned}$$

Thus

$$\{\mathbf{Var}_i[\mathbf{t}]\} = \{\mathbf{M}_i[\mathbf{t}^2] - \mathbf{M}_i[\mathbf{t}]^2\} = (2N-I)\tau - \tau_{sq}.$$

In our example,

$$\tau = N\xi = \begin{matrix}\mathbf{s}_2 \\ \mathbf{s}_3 \\ \mathbf{s}_4\end{matrix}\begin{pmatrix} 17/5 \\ 18/5 \\ 11/5 \end{pmatrix}$$

$$(2N-I) = \begin{pmatrix} 9/5 & 12/5 & 8/5 \\ 6/5 & 13/5 & 12/5 \\ 2/5 & 6/5 & 9/5 \end{pmatrix}$$

$$(2N-I)\tau = \begin{pmatrix} 457/25 \\ 468/25 \\ 241/25 \end{pmatrix} \qquad \tau_{sq} = \begin{pmatrix} 289/25 \\ 324/25 \\ 121/25 \end{pmatrix}$$

$$\tau_2 = \begin{matrix}\mathbf{s}_2 \\ \mathbf{s}_3 \\ \mathbf{s}_4\end{matrix}\begin{pmatrix} 168/25 \\ 144/25 \\ 120/25 \end{pmatrix}.$$

We see that one expects to reach the boundary most quickly from s_4. This is not surprising, since it is easier to reach the boundary from an outside state than from the middle, and it is more probable that the process moves to the right. But we again note that the variance is sizable.

We have computed means only for measures in which the process starts in a given state s_i. But it is easy to obtain from this the means and variances for an arbitrary initial probability vector.

3.3.6 COROLLARY. *If π is the initial probability vector for an absorbing chain, and π' consists of the last s components of π, i.e. π' gives the initial probabilities for the transient states, then*

$$\begin{aligned}\{\mathbf{M}_\pi[\mathbf{n}_j]\} &= \pi' N\\ \{\mathbf{Var}_\pi[\mathbf{n}_j]\} &= \pi' N(2N_{\mathrm{dg}} - I) - (\pi' N)_{\mathrm{sq}}\\ \{\mathbf{M}_\pi[\mathbf{t}]\} &= \pi'\tau\\ \{\mathbf{Var}_\pi[\mathbf{t}]\} &= \pi'(2N - I)\tau - (\pi'\tau)_{\mathrm{sq}}.\end{aligned}$$

PROOF. This is an immediate consequence of the fact that for any function $\mathbf{f}$, $\mathbf{M}_\pi[\mathbf{f}] = \pi \mathbf{M}_i[\mathbf{f}]$, which follows from the nature of the tree measure. The right sides contain π' rather than π, since the various means are 0 if the initial state is absorbing.

Our remaining applications will concern the question of which absorbing state is likely to capture the process.

3.3.7 THEOREM. *If b_{ij} is the probability that the process starting in transient state s_i ends up in absorbing state s_j, then*

$$\{b_{ij}\} = B = NR, \qquad s_i \in \mathbf{T}, \qquad s_j \in \tilde{\mathbf{T}}.$$

PROOF. Starting in s_i, the process may be captured in s_j in one or more steps. The probability of capture on a single step is p_{ij}. If this does not happen, the process may move either to another absorbing state (in which case it is impossible to reach s_j), or to a transient state s_k. In the latter case there is probability b_{kj} of being captured in the right state. Hence we have

$$b_{ij} = p_{ij} + \sum_{s_k \in \mathbf{T}} p_{ik} b_{kj},$$

which can be written in matrix form as

$$B = R + QB.$$

Thus

$$B = (I - Q)^{-1} R = NR.$$

An alternative proof is based on the following observation: Every time that the process is in transient state $\mathbf{s}_k$, it has probability p_{kj} of going to $\mathbf{s}_j$. Hence it is possible to show that

$$b_{ij} = \sum_{\mathbf{s}_k \in \mathbf{T}} \mathbf{M}_i[\mathbf{n}_k] \cdot p_{kj}.$$

This gives directly that

$$B = NR.$$

In our example

$$R = \begin{pmatrix} 1/3 & 0 \\ 0 & 0 \\ 0 & 2/3 \end{pmatrix} \qquad B = \mathbf{NR} = \begin{matrix} & \mathbf{s}_1 & \mathbf{s}_5 \\ \mathbf{s}_2 & 7/15 & 8/15 \\ \mathbf{s}_3 & 1/5 & 4/5 \\ \mathbf{s}_4 & 1/15 & 14/15 \end{matrix}.$$

It is worth noting that for each starting state the sum of the two absorption probabilities is 1. By Theorem 3.1.1 it will always be true that $NR\xi_{r-s} = \xi_s$. It is also easy to verify this directly.

The further to the right we start, the more probable it is, of course, that the process will end up at the right end. It is interesting to see that even in the leftmost transient state the probability is somewhat greater for capture on the right.

3.3.8 COROLLARY. *If ρ_a is the a-th column of R, i.e. $\rho_a = p_{ia}$ for $\mathbf{s}_i$ in* **T** *and for fixed a, then $N\rho_a$ gives the probabilities of absorption in the given absorbing state $\mathbf{s}_a$, for any transient state as initial state.*

This corollary is useful if we are interested in a single absorbing state.

3.3.9 THEOREM. *If B^* is the $r \times r$ matrix whose entry b^*_{ij} gives the probability of being absorbed in $\mathbf{s}_j$, starting in $\mathbf{s}_i$, for* all *states $\mathbf{s}_i$ and $\mathbf{s}_j$, then*

$$PB^* = B^*.$$

PROOF. If $s_j \in T$, then $b^*_{ij} = 0$. Hence the last s columns of B^* are 0. Consider $\mathbf{s}_j$ absorbing. If $\mathbf{s}_i \in \mathbf{T}$, then $b^*_{ij} = b_{ij}$, as in § 3.3.7. If $\mathbf{s}_i$ is also absorbing, then $b^*_{ij} = d_{ij}$. Hence we have

$$B^* = \left(\begin{array}{c|c} I & 0 \\ \hline B & 0 \end{array}\right)$$

$$PB^* = \left(\begin{array}{c|c} I & 0 \\ \hline R & Q \end{array}\right)\left(\begin{array}{c|c} I & 0 \\ \hline B & 0 \end{array}\right) = \left(\begin{array}{c|c} I & 0 \\ \hline R+QB & 0 \end{array}\right).$$

But $R + QB = R + QNR = R + (N-I)R = NR = B.$

Hence $$PB^* = B^*.$$

We thus see that the r-component column vector giving the probabilities of absorption in an absorbing state $\mathbf{s}_j$ is a fixed vector of P, and its first $r-s$ components are 0, except the j-th, which is 1. This determines the vector. This method of finding the absorption probabilities is useful if we are not interested in finding N.

In our example it is easily verified that

$$\begin{pmatrix} 1 \\ 0 \\ 7/15 \\ 1/5 \\ 1/15 \end{pmatrix} \quad \text{and} \quad \begin{pmatrix} 0 \\ 1 \\ 8/15 \\ 4/5 \\ 14/15 \end{pmatrix}$$

are fixed vectors of P.

We will now supply the missing step for Theorem 3.3.3.

3.3.10 THEOREM. *$\mathbf{M}_i[\mathbf{n}^2_j]$ is finite for any absorbing chain, and any $\mathbf{s}_i, \mathbf{s}_j \in \mathbf{T}$.*

PROOF.
$$\begin{aligned}\mathbf{M}_i[\mathbf{n}^2_j] &= \mathbf{M}_i\left[\left(\sum_{k=0}^{\infty} \mathbf{u}^k{}_j\right)^2\right] \\ &= \mathbf{M}_i\left[\sum_{k=0}^{\infty}\sum_{l=0}^{\infty} \mathbf{u}^k{}_j\mathbf{u}^l{}_j\right] \\ &= \sum_{k=0}^{\infty}\sum_{l=0}^{\infty} \mathbf{M}_i[\mathbf{u}^k{}_j\mathbf{u}^l{}_j].\end{aligned}$$

$\mathbf{M}_i[\mathbf{u}^k{}_j\mathbf{u}^l{}_j]$ is the probability that the process is in $\mathbf{s}_j$ both on step k and on l, starting in $\mathbf{s}_i$. If we let $m = \mathbf{min}(k, l)$, $d = |k-l|$, then this is the probability of being in $\mathbf{s}_j$ after m steps, and of returning d steps later. Hence $\mathbf{M}_i[\mathbf{u}^k{}_j\mathbf{u}^l{}_j] = p^{(m)}{}_{ij}p^{(d)}{}_{jj}$.

$$\begin{aligned}\mathbf{M}_i[\mathbf{n}^2_j] &= \sum_{k=0}^{\infty}\sum_{l=0}^{\infty} p^{(m)}{}_{ij}p^{(d)}{}_{jj} \\ &\leq \sum_{k=0}^{\infty}\sum_{l=0}^{\infty} (b\cdot c^m)(b\cdot c^d) \\ &= b^2\sum_{k=0}^{\infty}\sum_{l=0}^{\infty} c^n \quad \text{where } n = \mathbf{max}\,(k, l) \\ &= b^2\sum_{n=0}^{\infty}(2n+1)c^n, \quad \text{which is finite.}\end{aligned}$$

§ 3.4 Examples

EXAMPLE 3.4.1 (Example 1 of § 2.2 continued). In the random walk we find:

$$N = \frac{1}{p^2+q^2}\begin{pmatrix} p+q^2 & p & p^2 \\ q & 1 & p \\ q^2 & q & q+p^2 \end{pmatrix}$$

$$N_2 = \frac{pq}{(p^2+q^2)^2}\begin{pmatrix} p+q^2 & 1+2p & p+p^2 \\ 1 & 2 & 1 \\ q+q^2 & 1+2q & q+p^2 \end{pmatrix}$$

$$\tau = \frac{1}{p^2+q^2}\begin{pmatrix} 1+2p^2 \\ 2 \\ 1+2q^2 \end{pmatrix} \qquad \tau_2 = \frac{4pq}{(p^2+q^2)^2}\begin{pmatrix} 1+2p \\ 2 \\ 1+2q \end{pmatrix}$$

$$B = \frac{1}{p^2+q^2}\begin{pmatrix} pq+q^3 & p^3 \\ q^2 & p^2 \\ q^3 & pq+p^3 \end{pmatrix}.$$

In particular, if $p = 1/2$ (Example 1b), then

$$N = \begin{pmatrix} 3/2 & 1 & 1/2 \\ 1 & 2 & 1 \\ 1/2 & 1 & 3/2 \end{pmatrix} \qquad N_2 = \begin{pmatrix} 3/4 & 2 & 3/4 \\ 1 & 2 & 1 \\ 3/4 & 2 & 3/4 \end{pmatrix}$$

$$\tau = \begin{pmatrix} 3 \\ 4 \\ 3 \end{pmatrix} \qquad \tau_2 = \begin{pmatrix} 8 \\ 8 \\ 8 \end{pmatrix} \qquad B = \begin{pmatrix} 3/4 & 1/4 \\ 1/2 & 1/2 \\ 1/4 & 3/4 \end{pmatrix}.$$

And if $p = 1$ (Example 1c), then

$$N = \begin{pmatrix} 1 & 1 & 1 \\ 0 & 1 & 1 \\ 0 & 0 & 1 \end{pmatrix} \qquad \tau = \begin{pmatrix} 3 \\ 2 \\ 1 \end{pmatrix} \qquad B = \begin{pmatrix} 0 & 1 \\ 0 & 1 \\ 0 & 1 \end{pmatrix},$$

and the variances are all 0.

This last case is easily interpreted if we remember that the process in this case must move to the right.

EXAMPLE 3.4.2 (Example 10 of § 2.2 continued). In the college process we have, letting $t = \dfrac{r}{p+r}$:

$$N = \frac{1}{(p+r)}\begin{pmatrix}1 & 0 & 0 & 0\\ t & 1 & 0 & 0\\ t^2 & t & 1 & 0\\ t^3 & t^2 & t & 1\end{pmatrix}$$

$$N_2 = \frac{1}{(p+r)^2}\begin{pmatrix}q & 0 & 0 & 0\\ qt+t-t^2 & q & 0 & 0\\ qt^2+t^2-t^4 & qt+t-t^2 & q & 0\\ qt^3+t^3-t^6 & qt^2+t^2-t^4 & qt+t-t^2 & q\end{pmatrix}$$

$$\tau = \frac{1}{p}\begin{pmatrix}1-t\\ 1-t^2\\ 1-t^3\\ 1-t^4\end{pmatrix}$$

$$\tau_2 = \frac{1}{p(p+r)}\begin{pmatrix}q(1-t)\\ q(1-t^2)+t-2t^2+t^3\\ q(1-t^3)+t+t^2-4t^3+t^4+t^5\\ q(1-t^4)+t+t^2+t^3-6t^4+t^5+t^6+t^7\end{pmatrix}$$

$$B = \begin{pmatrix}1-t & t\\ 1-t^2 & t^2\\ 1-t^3 & t^3\\ 1-t^4 & t^4\end{pmatrix}.$$

The probability of graduating from each class depends only on the ratio $t = \dfrac{r}{p+r}$. This ratio is the conditional probability that the man is promoted rather than flunked out, given that he leaves his present class. Having successive powers of this ratio can be interpreted as saying that each time he leaves his class he must be promoted rather than flunked out, but it does not matter how long he stays in his present class. The formulas simplify greatly if we eliminate the possibility of having a man repeat the class, that is if $q=0$. In that case, $t = \dfrac{r}{p+r} = r$, and

$$N = \begin{pmatrix} 1 & 0 & 0 & 0 \\ r & 1 & 0 & 0 \\ r^2 & r & 1 & 0 \\ r^3 & r^2 & r & 1 \end{pmatrix} \qquad N_2 = \begin{pmatrix} 0 & 0 & 0 & 0 \\ r-r^2 & 0 & 0 & 0 \\ r^2-r^4 & r-r^2 & 0 & 0 \\ r^3-r^6 & r^2-r^4 & r-r^2 & 0 \end{pmatrix}$$

$$\tau = \begin{pmatrix} 1 \\ 1+r \\ 1+r+r^2 \\ 1+r+r^2+r^3 \end{pmatrix} \qquad \tau_2 = pr \begin{pmatrix} 0 \\ 1 \\ 1+3r+r^2 \\ 1+3r+6r^2+3r^3+r^4 \end{pmatrix}.$$

B is unchanged.

In the numerical Example 10a (cf. § 3.2.5) we have:

$$N = \begin{pmatrix} 1.11 & 0 & 0 & 0 \\ .86 & 1.11 & 0 & 0 \\ .67 & .86 & 1.11 & 0 \\ .52 & .67 & .86 & 1.11 \end{pmatrix};$$

$$N_2 = \begin{pmatrix} .12 & 0 & 0 & 0 \\ .31 & .12 & 0 & 0 \\ .37 & .31 & .12 & 0 \\ .37 & .37 & .31 & .12 \end{pmatrix}$$

$$\tau = \begin{pmatrix} 1.11 \\ 1.98 \\ 2.65 \\ 3.17 \end{pmatrix} \qquad \tau_2 = \begin{pmatrix} .12 \\ .43 \\ 1.13 \\ 2.22 \end{pmatrix}$$

	FLUNK OUT	GRADUATE	
$B =$	.22	.78	SENIOR
	.40	.60	JUNIOR
	.53	.47	SOPHOMORE
	.63	.37	FRESHMAN

Thus a student must reach the junior year before he has a better than even chance of graduating.

EXAMPLE 3.4.3 (Example 9 of § 2.2). In the urn example the five vectors and matrices are:

$$N = \begin{pmatrix} 1 & 1 & 1 \\ 0 & {}^4/_3 & {}^2/_3 \\ 0 & {}^2/_3 & {}^4/_3 \end{pmatrix} \qquad N_2 = \begin{pmatrix} 0 & {}^1/_3 & {}^2/_3 \\ 0 & {}^4/_9 & {}^2/_3 \\ 0 & {}^2/_3 & {}^4/_9 \end{pmatrix}$$

$$\tau = \begin{pmatrix} 3 \\ 2 \\ 2 \end{pmatrix} \qquad \tau_2 = \begin{pmatrix} 2 \\ 2 \\ 2 \end{pmatrix}$$

$$B = \begin{matrix} & \begin{matrix} s_1 & s_2 & s_3 \end{matrix} \\ \begin{matrix} s_4 \\ s_5 \\ s_6 \end{matrix} & \begin{pmatrix} {}^1/_2 & {}^1/_4 & {}^1/_4 \\ {}^1/_2 & {}^1/_3 & {}^1/_6 \\ {}^1/_2 & {}^1/_6 & {}^1/_3 \end{pmatrix} \end{matrix}.$$

Since the process must leave s_4 immediately and cannot return, there is 0 variance for the number of times in this state. Of the remaining variances the diagonal elements are smallest—this is due to the stabilizing effect of having to count the original position.

The B matrix needs special interpretation in this case. Since the states s_1, s_2, and s_3 were not absorbing in the original process, the "absorption probabilities" must be interpreted as probabilities of entering the ergodic set at the given state. Thus, for example, if the process starts with both balls unpainted (state s_4), then there is probability ${}^1/_2$ that the first time both balls are painted there will be one of each color, ${}^1/_4$ that they will both be red, and ${}^1/_4$ that they will both be black. It should be noted that these probabilities are the same as if we had assigned the two balls colors independently and at random.

§ 3.5 Extension of results.

We will see that results obtained in § 3.3 can be applied to a wider variety of problems.

3.5.1 DEFINITION. *A set* S *of states is an* open set *if from every state in* S *it is possible to go to a state in* $\tilde{\mathrm{S}}$.

It is easy to think of examples of open sets: A set consisting of a single state is open (unless the state is absorbing), so is a set of transient states, so is a proper subset of an ergodic set, etc. The following theorem characterizes these sets.

3.5.2 THEOREM. *A set* S *of states is open if and only if no ergodic set is a subset of* S.

PROOF. If an ergodic set is contained in S, then there is no escape from this set once it is entered; hence S is not open.

On the other hand we know that from every state we can reach an ergodic state. And from an ergodic state we can reach all the elements of its ergodic set. Hence if there is no ergodic set contained in S, then for every element of S we can find an ergodic state in $\tilde{\mathrm{S}}$ which can be reached from the given state. Hence S is open.

3.5.3 THEOREM. *If* S *is an open set of states, and all the states in* $\tilde{\mathrm{S}}$ *are made absorbing states, then the resulting Markov chain is absorbing, and its transient states are the elements of* S.

PROOF. Since S is open, from every state of it we can reach a state in $\tilde{\mathrm{S}}$—which must be an absorbing state. Hence the chain is absorbing. And since from each element of S we can reach an absorbing state, the elements of S must all be transient states in the new process.

3.5.4 THEOREM. *Let* S *be an open set of s states. Let Q be the $s \times s$ submatrix of P corresponding to these states. Let ρ_a be the s-component column vector with components p_{ia}, where the s_i are the elements of* S *and $\mathrm{s}_a \in \tilde{\mathrm{S}}$. Let the process start in s_i. Then:*

(1) *The ij-component of $N = (I - Q)^{-1}$ is the mean number of times the process is in s_j before leaving* S.

(2) *The ij-components of $N_2 = N(2N_{dg} - I) - N_{sq}$ is the variance of the same function.*

(3) *The i-th component of $\tau = N\xi$ is the mean number of steps needed to leave* S.

(4) *The i-th component of $\tau_2 = (2N - I)\tau - \tau_{sq}$ is the variance of the same function.*

(5) *The i-th component of $N\rho_a$ is the probability that the process goes to s_a when it leaves* S.

PROOF. The various parts of this theorem are a direct consequence of the corresponding results in § 3.3, due to Theorem 3.5.3.

As an application of this theorem consider the following problem. Let s_j and s_k be any two states in a regular Markov chain. Assume that the process is started at a third state. What is the probability of reaching s_k before s_j? This probability may be found from 3.5.4(5) by choosing S to be the set of all states in the ergodic set except s_j and s_k.

3.5.5 EXAMPLE. Consider the random walk Example 6 of Chapter II. The transition matrix is

$$P = \begin{matrix} s_1 \\ s_2 \\ s_3 \\ s_4 \\ s_5 \end{matrix} \begin{pmatrix} 0 & 1/4 & 1/4 & 1/4 & 1/4 \\ 1/3 & 1/3 & 1/3 & 0 & 0 \\ 0 & 1/3 & 1/3 & 1/3 & 0 \\ 0 & 0 & 1/3 & 1/3 & 1/3 \\ 1/4 & 1/4 & 1/4 & 1/4 & 0 \end{pmatrix},$$

and since from any state we can move to any other state in two steps, the Markov chain is regular. Hence any proper subset of the states is open. Let S consist of the last three states.

$$Q = \begin{matrix} s_3 \\ s_4 \\ s_5 \end{matrix} \begin{pmatrix} 1/3 & 1/3 & 0 \\ 1/3 & 1/3 & 1/3 \\ 1/4 & 1/4 & 0 \end{pmatrix}$$

$$N = \begin{pmatrix} 21/9 & 12/9 & 4/9 \\ 15/9 & 24/9 & 8/9 \\ 9/9 & 9/9 & 12/9 \end{pmatrix}$$

$$N_2 = \begin{pmatrix} 252/81 & 324/81 & 44/81 \\ 270/81 & 360/81 & 56/81 \\ 216/81 & 270/81 & 36/81 \end{pmatrix}$$

$$\tau = \begin{pmatrix} 37/9 \\ 47/9 \\ 30/9 \end{pmatrix} \qquad \tau_2 = \begin{pmatrix} 1220/81 \\ 1114/81 \\ 1062/81 \end{pmatrix}$$

$$N\rho_1 = \begin{pmatrix} 1/9 \\ 2/9 \\ 3/9 \end{pmatrix}.$$

The N matrix tells us the mean number of times that the process is in each of the last three states, before it goes to one of the first two states. We see that the numbers are small if the process starts in the last state. But this is intuitively clear, since in this case it has a $1/2$ probability of "escaping" on the first step. For the same reason, the mean number of times that it is in the last state is small, no matter where the process starts. However, from N_2 we see that the former numbers have much greater variances than the latter.

From τ we see that it takes longest to escape from s_4, which has no connection to $\tilde{\mathbf{S}}$. Indeed, the differences in mean number of steps to escape can be accounted for by the number of connections the three states have with outside states. Note that while the means differ considerably, the variances are roughly the same.

Finally, the vector $\mathrm{N}\rho_1$ gives us the "exit probabilities" for state $\mathbf{s}_1$, i.e. the probabilities (depending on starting state) of going to $\mathbf{s}_1$ when the process leaves S; or, stated otherwise, the probability of hitting $\mathbf{s}_1$ before hitting $\mathbf{s}_2$. These probabilities seem to depend very simply on the number of steps necessary to reach $\mathbf{s}_1$ from the starting state (going through S).

3.5.6 THEOREM. *Let $\mathbf{r}_i$ be the function giving the number of times that the process remains in the non-absorbing state $\mathbf{s}_i$ once the state is entered (including the entering step). Then*

$$\mathbf{M}_i[\mathbf{r}_i] = 1/(1-p_{ii}), \tag{a}$$

$$\mathbf{Var}_i[\mathbf{r}_i] = p_{ii}/(1-p_{ii})^2. \tag{b}$$

And the conditional probability of the process going to $\mathbf{s}_j$, given that it leaves $\mathbf{s}_i$, is

$$p_{ij}/(1-p_{ii}). \tag{c}$$

PROOF. The set whose only element is $\mathbf{s}_i$ is an open set. We apply Theorem 3.5.4 to this set. In this case N is a 1×1 matrix, and hence identical with τ; its only component is $1/(1-p_{ii})$. Hence (a) is a consequence of either (1) or (3) of Theorem 3.5.4. Similarly, $N_2 = \tau_2$, and (b) is a consequence of either (2) or (4) of Theorem 3.5.4. We obtain (c) from 3.5.4(5) by choosing the vector ρ_j whose only component is p_{ij}. Since $\mathbf{s}_i$ is not absorbing, $p_{ii} < 1$, hence our quantities are well defined.

One type of concept that we have not investigated as yet is illustrated by the question of whether the process ever enters a given transient state. This and related questions are taken up in Theorems 3.5.7, 3.5.8, and 3.5.9. For these theorems we will let $\mathbf{n}_j$ be the number of times that the process is in transient state $\mathbf{s}_j$, $\mathbf{m}$ be the total number of transient states it will ever be in, and h_{ij} be the probability that the process will ever go to transient state $\mathbf{s}_j$, starting in transient state $\mathbf{s}_i$ (not counting the initial state).

3.5.7 THEOREM.

$$H = \{h_{ij}\} = (N-I)N_{\mathrm{dg}}^{-1}.$$

PROOF.

$$\{\mathbf{M}_i[\mathbf{n}_j]\} = \{d_{ij}\} + \{h_{ij}\mathbf{M}_j[\mathbf{n}_j]\}$$

or

$$\{n_{ij}\} = I + \{h_{ij}n_{jj}\}$$

or

$$N = I + HN_{\mathrm{dg}}.$$

Hence

$$H = (N-I)N_{\mathrm{dg}}^{-1}.$$

3.5.8 THEOREM. $\{\mathbf{Pr}_i[\mathbf{n}_j - \mathbf{d}_{ij} = k]\} =$

$$\begin{cases} E-H & \text{if } k = 0 \\ H \cdot H_{\mathrm{dg}}^{k-1}[I - H_{\mathrm{dg}}] = (N-I)N_{\mathrm{dg}}^{-2}(I - N_{\mathrm{dg}}^{-1})^{k-1} & \text{if } k > 0 \end{cases}$$

This theorem determines the probability of going to a given transient state exactly k times. The theorem is an immediate consequence of the following consideration: To go to a given state k times one must go there at least once, then one must return $k-1$ times, and one must not return again.

3.5.9 THEOREM.

$$\mu = \{\mathbf{M}_i[\mathbf{m}]\} = [H + (I - H_{\mathrm{dg}})]\xi = NN_{\mathrm{dg}}^{-1}\xi.$$

PROOF. The mean number of transient states occupied is equal to the sum of the probabilities of ever being in the various states. If the process starts in $\mathbf{s}_i$, the probability of ever being in $\mathbf{s}_j$ is h_{ij} if $i \neq j$, and is 1 if $i = j$.

If we apply Theorem 3.5.7 to Example 1, we obtain

$$H = \begin{pmatrix} \dfrac{pq}{1-pq} & p & \dfrac{p^2}{1-pq} \\ \\ \dfrac{q}{1-pq} & 2pq & \dfrac{p}{1-pq} \\ \\ \dfrac{q^2}{1-pq} & q & \dfrac{pq}{1-pq} \end{pmatrix}.$$

We see, for example, that if $q = 0$, then all entries on and below the main diagonal are 0. This means that if the process is sure to move to the right, then it can never re-enter the starting state, nor can it enter a state to the left of the starting state.

$$\mu = \frac{1}{1-pq}\begin{pmatrix} 1+p^2+p^3 \\ 2-pq \\ 1+q^2+q^3 \end{pmatrix}.$$

If $q=0$ the vector $\mu=\begin{pmatrix}3\\2\\1\end{pmatrix}$, which is obvious in this case since it moves directly to the right boundary, passing through the intermediate states only once.

3.5.10 THEOREM. *The mean and variance of the number of changes of state in an absorbing chain can be calculated by setting $p_{ii}=0$ for all transient states, and dividing each row by its row-sum. The i-th component of the new τ gives the mean number of changes of state for the original process. The variance of the same function is given by the new τ_2.*

PROOF. Assume that the Markov chain is started in a non-absorbing state. We form a new process in which the n-th outcome function is defined as follows: If the original chain is absorbed at state s_k before making n changes of state, then $\hat{f}_n=s_k$. If not, $\hat{f}_n$ is the state to which the process moved on the n-th change of state. The new process is clearly a Markov chain. The transition probabilities are the same as P for s_i absorbing. For s_i non-absorbing

$$\hat{p}_{ii} = \mathbf{Pr}_i[\hat{f}_1=i] = 0$$

$$\hat{p}_{ij} = \mathbf{Pr}_i[\hat{f}_1=j] = \sum_{n=0}^{\infty} p^{(n)}{}_{ii}p_{ij} = \frac{p_{ij}}{1-p_{ii}}.$$

From this new transition matrix we can obtain the mean and variance of the time to absorption for the process $\hat{f}_1, \hat{f}_2, \ldots$. This time represents the number of changes of state in the original chain started in state s_i.

We can also find the mean number of times that the process does not change its state while it is among the transient states. This is found by taking the mean number of times to reach the absorbing states and subtracting the mean number of changes of state.

If we want to illustrate Theorem 3.5.10 by the college example, Example 10 of § 2.2, we set $p_{ii}=0$, $i=3, 4, 5, 6$, and renormalize:

$$\hat{P} = \left(\begin{array}{cc|cccc} 1 & 0 & 0 & 0 & 0 & 0 \\ 0 & 1 & 0 & 0 & 0 & 0 \\ \hline .22 & .78 & 0 & 0 & 0 & 0 \\ .22 & 0 & .78 & 0 & 0 & 0 \\ .22 & 0 & 0 & .78 & 0 & 0 \\ .22 & 0 & 0 & 0 & .78 & 0 \end{array}\right);$$

$$N = \begin{pmatrix} 1 & 0 & 0 & 0 \\ .78 & 1 & 0 & 0 \\ .61 & .78 & 1 & 0 \\ .47 & .61 & .78 & 1 \end{pmatrix}$$

$$\tau = \begin{pmatrix} 1.00 \\ 1.78 \\ 2.38 \\ 2.85 \end{pmatrix} \qquad \tau_2 = \begin{pmatrix} 0 \\ .17 \\ .68 \\ 1.51 \end{pmatrix}$$

By comparing these results with Example 3.4.2, we note that the mean number of steps to absorption is somewhat higher than the mean number of changes of state (but not by much, since repetition of a state is rare), and that the variance of the former is considerably higher than that of the latter.

Another interesting use of conditional probabilities for absorbing chains is the following. Assume that for an absorbing chain we start in a non-absorbing state and compute all probabilities relative to the hypothesis that the process ends up in a given absorbing state, say $\mathbf{s}_1$. Then we obtain a new absorbing chain with a single absorbing state $\mathbf{s}_1$. The non-absorbing states will be as before, except that we have new transition probabilities. We compute these as follows. Let $\mathbf{p}$ be the statement "the original process is absorbed in state $\mathbf{s}_1$." Then if $\mathbf{s}_i$ is a non-absorbing state, the transition probabilities for the new process are

$$\mathbf{Pr}_i[\mathbf{f}_1 = \mathbf{s}_j | \mathbf{p}] = \frac{\mathbf{Pr}_i[\mathbf{f}_1 = \mathbf{s}_j \wedge \mathbf{p}]}{\mathbf{Pr}_i[\mathbf{p}]} = \frac{\mathbf{Pr}_i[\mathbf{p} | \mathbf{f}_1 = \mathbf{s}_j] \cdot \mathbf{Pr}_i[\mathbf{f}_1 = \mathbf{s}_j]}{\mathbf{Pr}_i[\mathbf{p}]}.$$

$$\hat{p}_{ij} = \frac{b_{j1} p_{ij}}{b_{i1}}.$$

This formula applies for $j = 1$ if we interpret $b_{11} = 1$. The standard form for $\hat{P}$ may be obtained as follows. The matrix $\hat{R}$ is a column vector with $\hat{R} = \left\{\frac{p_{i1}}{b_{i1}}\right\}$. Let D_0 be a diagonal matrix with diagonal entries b_{j1}, for $\mathbf{s}_j$ non-absorbing. Then

$$\hat{Q} = D^{-1}{}_0 Q D_0.$$

From this we see that

$$\hat{Q}^n = D^{-1}{}_0 Q^n D_0$$

and

$$\hat{N} = D^{-1}{}_0[I + Q + Q^2 + \cdots]D_0$$
$$= D^{-1}{}_0 N D_0.$$

$\hat{B} = \xi$ and $\hat{\tau}$ may be obtained from $\hat{N}$.

EXAMPLE. Consider Example 1a, § 3.2.5. Let us consider the process obtained by assuming that the original chain is absorbed in state s_1. Then the new matrix $\hat{Q}$ is

$$\hat{Q} = \begin{pmatrix} 15/7 & 0 & 0 \\ 0 & 15/3 & 0 \\ 0 & 0 & 15/1 \end{pmatrix} \overset{\begin{matrix} s_2 & s_3 & s_4 \end{matrix}}{\begin{pmatrix} 0 & 2/3 & 0 \\ 1/3 & 0 & 2/3 \\ 0 & 1/3 & 0 \end{pmatrix}} \begin{pmatrix} 7/15 & 0 & 0 \\ 0 & 3/15 & 0 \\ 0 & 0 & 1/15 \end{pmatrix}$$

$$= \begin{pmatrix} 0 & 2/7 & 0 \\ 7/9 & 0 & 2/9 \\ 0 & 1 & 0 \end{pmatrix}$$

so that

$$\hat{P} = \begin{matrix} & \begin{matrix} s_1 & s_2 & s_3 & s_4 \end{matrix} \\ \begin{matrix} s_1 \\ s_2 \\ s_3 \\ s_4 \end{matrix} & \begin{pmatrix} 1 & 0 & 0 & 0 \\ 5/7 & 0 & 2/7 & 0 \\ 0 & 7/9 & 0 & 2/9 \\ 0 & 0 & 1 & 0 \end{pmatrix} \end{matrix}$$

$$\hat{N} = \begin{pmatrix} 15/7 & 0 & 0 \\ 0 & 15/3 & 0 \\ 0 & 0 & 15 \end{pmatrix} \begin{pmatrix} 7/5 & 6/5 & 4/5 \\ 3/5 & 9/5 & 6/5 \\ 1/5 & 3/5 & 7/5 \end{pmatrix} \begin{pmatrix} 7/15 & 0 & 0 \\ 0 & 3/15 & 0 \\ 0 & 0 & 1/15 \end{pmatrix}$$

$$= \begin{pmatrix} 7/5 & 18/35 & 4/35 \\ 7/5 & 9/5 & 2/5 \\ 7/5 & 9/5 & 7/5 \end{pmatrix}$$

$$\hat{\tau} = \begin{pmatrix} 71/35 \\ 18/5 \\ 23/5 \end{pmatrix}.$$

Exercises for Chapter III

For § 3.1

1. Put the following matrices in the canonical form for absorbing chains.

$$\text{(a)} \qquad P = \begin{array}{c} \\ s_1 \\ s_2 \\ s_3 \end{array} \begin{array}{c} \begin{array}{ccc} s_1 & s_2 & s_3 \end{array} \\ \begin{pmatrix} 1/3 & 1/3 & 1/3 \\ 0 & 1 & 0 \\ 1/3 & 1/2 & 1/6 \end{pmatrix} \end{array}$$

$$\text{(b)} \qquad P = \begin{array}{c} \\ s_1 \\ s_2 \\ s_3 \\ s_4 \end{array} \begin{array}{c} \begin{array}{cccc} s_1 & s_2 & s_3 & s_4 \end{array} \\ \begin{pmatrix} 1 & 0 & 0 & 0 \\ 1 & 0 & 0 & 0 \\ 1/4 & 1/4 & 1/4 & 1/4 \\ 0 & 0 & 0 & 1 \end{pmatrix} \end{array}.$$

2. Apply Theorem 3.1.1 to an absorbing chain with a single absorbing state.

3. Apply the result of the previous exercise to an ergodic chain in which one state has been made absorbing. (See Chapter II, Exercise 21.)

4. In Example 8 of § 2.2 make state **R** into an absorbing state. What does Theorem 3.1.1, applied to the resulting absorbing chain, say about the weather in the Land of Oz? (That is, what do we learn about the *original* chain?)

For § 3.2

5. Compute the fundamental matrix for the absorbing chain with transition matrix.

$$P = \begin{array}{c} \\ s_1 \\ s_2 \\ s_3 \end{array} \begin{array}{c} \begin{array}{ccc} s_1 & s_2 & s_3 \end{array} \\ \begin{pmatrix} 1 & 0 & 0 \\ 0 & 1/2 & 1/2 \\ 1/2 & 0 & 1/2 \end{pmatrix} \end{array}$$

6. Compute the fundamental matrix for Example 11 of Chapter II when $c=0$, and $d \neq 0$.

7. Make Example 9 of § 2.2 into an absorbing chain by making all of the ergodic states absorbing. Find the fundamental matrix and interpret the entries of the first row of this matrix.

8. Show that if the fundamental matrix N is given for an absorbing chain, then N^{-1} exists and $Q = I - N^{-1}$.

9. Prove that $NQ = N - I$.

10. Check the results of Exercise 9, above, in Example 9 of § 2.2.

For § 3.3

11. If an absorbing chain has only one absorbing state, what can be said about the matrix B? In Example 8 of § 2.2 make **R** an absorbing state, compute N and B, and verify your statement.

12. Change Example 7 of § 2.2 into an absorbing chain by assuming that the process is stopped if a 0 or 9 is reached. Construct the new transition matrix, in canonical form.

13. In the example of Exercise 12, above, compute N, N_2, B, τ, τ_2.

14. In Example 8 of § 2.2 make **N** into an absorbing state. Compute the fundamental matrix for the resulting Markov chain. Find N_2, B, τ, τ_2. Interpret the results in terms of the original chain.

15. Compute N for the tank duel (Exercise 2 of Chapter II). From this find the mean length of the duel and the probability of each possible ending.

16. Carry out the computations of Exercise 15, above, for the modified tank duel (Exercise 3 of Chapter II). Which duel is more favorable to tank **A**?

17. In Example 10a (of § 3.2.5) find the probabilities of graduation by the method resulting from Theorem 3.3.9, that is, by finding a certain fixed column vector for the transition matrix.

18. The chain of Example 1a (cf. § 3.2.5) is started by means of a random device which make all five states equally likely as starting states. Find the means and variances of the number of times in the various transient states, and of the number of steps to absorption.

For § 3.5

19. In Example 9 of § 2.2, assume that initially both balls are unpainted. Find the mean number of draws before the first time that both balls are painted. When this occurs, what is the probability that both balls are red?

20. It is snowing in the Land of Oz today. Find the mean number of changes of weather that will occur before the next rainy day. Find the probability that there is at least one nice day before a rainy day.

21. For Example 1 of § 2.2 with $p = 1/2$, assume that it is known that the process is absorbed in state $\mathbf{s}_1$. Find the transition matrix for the new conditional process. Find the mean time to absorption.

22. Compute the following quantities for the tank duel (see Exercise 2 of Chapter II).

(a) The mean and variance of the number of rounds for which all three tanks remain active.
(b) The probability that at some stage **A** and **C** will still be active, but **B** is no longer active.
(c) The probability that at some stage **A** and **B** will still be active, but **C** is no longer active.
(d) The probability that **A** and **B** will be eliminated on the same round.
(e) $\hat{P}$, $\hat{N}$, $\hat{\tau}$, assuming that **C** wins the duel.
(f) $\hat{P}$, $\hat{N}$, $\hat{\tau}$, assuming that no tank survives.

23. In the tank duel (Exercise 2 of Chapter II) let tank **A** have probability $^3/_4$ of hitting, tank **B** probability $^3/_5$, and tank **C** an unspecified probability p (with $p < ^3/_5$).

(a) Set up the transition matrix.
(b) Find the probability that tank **C** is the survivor.
(c) In the answer obtained in (b), let p tend to 0.
Interpret your result.

For the entire chapter

24. Seven boys are playing with a ball.
The first boy always throws it to the second boy.
The second boy is equally likely to throw it to the third or the seventh.
The third boy keeps the ball if he gets it.
The fourth boy always throws it to the sixth.
The fifth boy is equally likely to throw it to the fourth, sixth, or seventh boy.
The sixth boy always throws it to the fourth.
The seventh boy is equally likely to throw it to the first or fourth boy.

(a) Set up the transition matrix P.
(b) Classify the states.
(c) Put P into canonical form.
(d) Give an interpretation for the chain ending up in one of the ergodic sets.
(e) The ball is given to the fifth boy. Find the mean and variance of the number of times that the seventh boy has the ball, and find the mean and variance of the time to reach an ergodic set.

25. Given an absorbing Markov chain, we play a game as follows:

We start in a specified state, and carry the chain out till it reaches an absorbing state. If we reach s_a, we receive a payment of c_a. Form the column vector γ whose i-th component is the mean of the payment if we start in s_i.

(a) Prove that $P\gamma = \gamma$.
(b) Prove that for absorbing state s_a the a-th component of γ is c_a.
(c) Prove that these two conditions determine γ. (HINT: Consider the limit of $P^n\gamma$.)
(d) Let γ_a be the vector giving the probabilities of absorption in s_a. Show that γ can be expressed in terms of the γ_a.

CHAPTER IV

REGULAR MARKOV CHAINS

§ 4.1 **Basic theorems.** In this section we shall study the behavior of a regular Markov chain. We recall that a regular Markov chain is one that has no transient sets, and has a single ergodic set with only one cyclic class.

4.1.1 DEFINITION. *The transition matrix for a regular Markov chain is called a* regular transition matrix.

4.1.2 THEOREM. *A transition matrix is regular if and only if for some N, P^N has no zero entries.*

It was shown in Chapter II that a Markov chain was regular if and only if it is possible to be in any state after some number N of steps, no matter what the starting state. That is, if and only if P^N has no zero entries for some N.

4.1.3 THEOREM. *Let P be an $r \times r$ transition matrix having no zero entries. Let ϵ be the smallest entry of P. Let x be any r-component column vector, having maximum component M_0 and minimum component m_0, and let M_1 and m_1 be the maximum and minimum components for the vector Px. Then $M_1 \leqslant M_0$, $m_1 \geqslant m_0$, and*

$$M_1 - m_1 \leqslant (1-2\epsilon)(M_0 - m_0).$$

PROOF. Let x' be the vector obtained from x by replacing all components, except one m_0 component, by M_0. Then $x \leqslant x'$. Each component of Px' is of the form

$$a \cdot m_0 + (1-a) \cdot M_0 = M_0 - a(M_0 - m_0)$$

where $a \geqslant \epsilon$. Thus each such component is $\leqslant M_0 - \epsilon(M_0 - m_0)$. But since $x \leqslant x'$, we have

$$M_1 \leqslant M_0 - \epsilon(M_0 - m_0). \tag{1}$$

If we apply this result to the vector $-x$ we obtain

$$-m_1 \leqslant -m_0 - \epsilon(-m_0 + M_0). \tag{2}$$

Adding (1) and (2) we have

$$\begin{aligned} M_1 - m_1 &\leqslant M_0 - m_0 - 2\epsilon(M_0 - m_0) \\ &= (1-2\epsilon)(M_0 - m_0). \end{aligned}$$

This theorem gives us a simple proof of the following fundamental theorem for regular Markov chains.

4.1.4 THEOREM. *If P is a regular transition matrix then*

(i) *The powers P^n approach a probability matrix A.*
(ii) *Each row of A is the same probability vector $\alpha = \{a_1, a_2, \ldots, a_n\}$, that is $A = \xi\alpha$.*
(iii) *The components of α are positive.*

PROOF. We shall first assume that P has no zeros. Let ϵ be the minimum entry. Let ρ_j be a column vector with a 1 in the j-th component and 0 otherwise. Let M_n and m_n be the maximum and minimum components of the vector $P^n\rho_j$. Since $P^n\rho_j = P \cdot P^{n-1}\rho_j$, we have, from Theorem 4.1.3, that $M_1 \geqslant M_2 \geqslant M_3 \geqslant \cdots$ and $m_1 \leqslant m_2 \leqslant m_3 \leqslant \cdots$ and

$$M_n - m_n \leqslant (1-2\epsilon)(M_{n-1} - m_{n-1})$$

for $n \geqslant 1$. If we let $d_n = M_n - m_n$, this tells us that

$$d_n \leqslant (1-2\epsilon)^n \cdot d_0 = (1-2\epsilon)^n.$$

Thus as n tends to infinity d_n goes to 0, M_n and m_n approach a common limit, and therefore $P^n\rho_j$ tends to a vector with all components the same. Let a_j be this common value. It is clear that, for all n, $m_n \leqslant a_j \leqslant M_n$. In particular, since $0 < m_1$ and $M_1 < 1$, we have that $0 < a_j < 1$. Now $P^n\rho_j$ is the j-th column of P^n. Thus the j-th column of P^n tends to a vector with all components the same value a_j. That is, P^n tends to a matrix A with all rows the same vector $\alpha = \{a_1, a_2, \ldots, a_r\}$. Since the row-sums of P^n are always 1, the same must be true of the limit. This completes the proof for the case where the matrix has all positive entries.

Consider next the case that P is only assumed to be regular. Let N be such that P^N has no zero entries. Let ϵ' be the smallest entry of P^N. Applying the first part of the proof to the matrix P^N, we have

$$d_{kN} \leqslant (1-2\epsilon')^k \tag{3}$$

Therefore the sequence d_n, which is non-increasing, has a subsequence tending to 0. Thus d_n tends to zero and the rest of the proof is the same as in the proof for all positive entries.

4.1.5 COROLLARY. *Let P be a regular transition matrix. Let $a_j = \lim_{n\to\infty} p^{(n)}{}_{ij}$. Then there are constants b and r with $0 < r < 1$ such that*

$$p^{(n)}{}_{ij} = a_j + e^{(n)}{}_{ij}$$

with $|e^{(n)}{}_{ij}| \leqslant br^n$.

PROOF. We know that $|e^{(n)}{}_{ij}| \leqslant d_n$. Let N be such that P^N has no zero entries. Let ϵ be the smallest entry of P^N. Choose $r = (1-2\epsilon)^{1/N}$ and $b = 1/(1-2\epsilon) = r^{-N}$. If $n = kN$, then from (3), $d_n \leqslant r^n$. If $n = kN + n_1$, where $0 \leqslant n_1 \leqslant N$, then since d_n is non-increasing, $d_n \leqslant r^{n-n_1} \leqslant r^n \cdot r^{-N} = br^n$. The bound here obtained for $e^{(n)}{}_{ij}$ is useful for proving theorems, but it is very conservative as an estimate for the rate of convergence of $p^{(n)}{}_{ij}$.

4.1.6 THEOREM. *If P is a regular transition matrix and A and α are as given in Theorem 4.1.4, then*

(a) *For any probability vector π, $\pi \cdot P^n$ approaches the vector α as as n tends to infinity.*
(b) *The vector α is the unique probability vector such that $\alpha P = \alpha$.*
(c) $PA = AP = A$.

PROOF. If π is a probability vector, then $\pi\xi = 1$; hence $\pi A = \pi\xi\alpha = \alpha$. But $\pi \cdot P^n$ approaches $\pi \cdot A$. Hence it approaches α. This proves part (a).

Since the powers of P approach A, $P^{n+1} = P^n \cdot P$ approaches A, but it also approaches AP; hence $AP = A$. Similarly $PA = A$, proving (c). Any one row of this matrix equation states that $\alpha P = \alpha$. We now show that α is unique. Let β be any probability vector such that $\beta P = \beta$. By (a), $\beta \cdot P^n$ approaches α. But since $\beta P = \beta$, $\beta P^n = \beta$. Hence $\alpha = \beta$. Thus we have proved (b).

The matrix A and vector α will be referred to as the limiting matrix and limiting vector for the Markov chain determined by P.

Theorem 4.1.6 shows that for a regular transition matrix there is a row vector α which remains "fixed" when multiplied by P. Any other vector α' such that $\alpha' P = \alpha'$ is proportional to the probability vector α. The following theorem shows that any fixed column vector for P is proportional to ξ.

4.1.7 THEOREM. *If P is a regular transition matrix and $\rho = \{r_i\}$ is a column vector such that*

$$P\rho = \rho$$

then $\rho = c \cdot \xi$ for some constant c.

PROOF. Since $P\rho=\rho$, $P^2\rho=P\rho=\rho$ and in general $P^n\rho=\rho$. Hence also $A\rho=\rho$. Thus $r_i=\alpha\rho$. But this states that all components of ρ have the same value. That is $\rho=c\xi$ for some constant c.

In Chapter II we saw that if the process is started in each of the states with probabilities given by π, then the probabilities for being in each of the states after n steps are given by πP^n. For large n Theorem 4.1.6 states that πP^n is approximately α. Since α depends only on P and not on π, this may be described by saying that, for a regular Markov chain, the long range predictions are independent of the initial vector. Let us illustrate this in terms of Example 8 of Chapter II. The transition matrix for this example is

$$\begin{array}{c} \\ \mathbf{R} \\ \mathbf{N} \\ \mathbf{S} \end{array}\begin{array}{c} \begin{array}{ccc} \mathbf{R} & \mathbf{N} & \mathbf{S} \end{array} \\ \begin{pmatrix} 1/2 & 1/4 & 1/4 \\ 1/2 & 0 & 1/2 \\ 1/4 & 1/4 & 1/2 \end{pmatrix} \end{array}.$$

To find the vector $\alpha=(a_1, a_2, a_3)$, we must find a probability vector such that $\alpha P=\alpha$. That is, we must satisfy the following set of equations:

$$\begin{aligned} 1 &= a_1 + a_2 + a_3 \\ a_1 &= {}^1/_2 a_1 + {}^1/_2 a_2 + {}^1/_4 a_3 \\ a_2 &= {}^1/_4 a_1 + {}^1/_4 a_3 \\ a_3 &= {}^1/_4 a_1 + {}^1/_2 a_2 + {}^1/_2 a_3. \end{aligned}$$

The unique solution to these equations is

$$\alpha = ({}^2/_5, {}^1/_5, {}^2/_5).$$

The limit matrix A is then

$$A = \begin{pmatrix} .4 & .2 & .4 \\ .4 & .2 & .4 \\ .4 & .2 & .4 \end{pmatrix}.$$

Corollary 4.1.5 states that this limit is reached geometrically. This being a very fast kind of convergence, we would expect that, even for moderately large values of n, P^n should be approximated by A. The matrix P^5 is

$$P^5 = \begin{array}{c} \\ \mathbf{R} \\ \mathbf{N} \\ \mathbf{S} \end{array}\begin{array}{c} \begin{array}{ccc} \mathbf{R} & \mathbf{N} & \mathbf{S} \end{array} \\ \begin{pmatrix} .4004 & .2002 & .3994 \\ .4004 & .1992 & .4004 \\ .3994 & .2002 & .4004 \end{pmatrix} \end{array}.$$

Each row of P^5 gives the probability of each kind of weather five days after a particular kind of day. For example, the first row gives the probabilities for each kind of weather five days after a rainy day. The fact that the rows are so nearly equal means that today's weather in the Land of Oz may be considered to have very little effect on our predictions for five days from now.

§ 4.2 Law of large numbers for regular Markov chains. As we have seen in § 4.1, for a regular Markov chain there is a limiting probability a_j of being in state s_j independent of the starting state. In this section we shall prove that a_j also represents the fraction of the time that the process can be expected to be in state s_j for a large number of steps. This result will also be independent of the starting state.

To state the above result precisely, we shall need to introduce some new functions. Let $\mathbf{u}^{(n)}{}_j$ be a function with domain the tree $\mathbf{U}_n$ and with value 1 if the n-th step was to state s_j and 0 otherwise. We define $\mathbf{y}^{(n)}{}_j = \sum_{k=1}^{n} \mathbf{u}^{(k)}{}_j$. Then $\mathbf{y}^{(n)}{}_j$ is again a function with domain the tree $\mathbf{U}_n$ and value the number of times (not counting the initial position) that the process is in state s_j during the first n steps. The function $\mathbf{v}^{(n)}{}_j = \mathbf{y}^{(n)}{}_j/n$ gives the fraction of times in the first n steps that the process moves to state s_j.

4.2.1 THEOREM (*The Law of Large Numbers*). *Consider a regular Markov chain with limiting vector* $\alpha = (a_1, a_2, \ldots, a_r)$. *For any initial vector* π,

$$\mathbf{M}_\pi[\mathbf{v}^{(n)}{}_j] \to a_j \tag{a}$$

and for any $\epsilon > 0$

$$\mathbf{Pr}_\pi[|\mathbf{v}^{(n)}{}_j - a_j| > \epsilon] \to 0 \tag{b}$$

as n tends to infinity.

PROOF. According to Theorem 1.8.10, to prove this theorem it is sufficient to prove that $\mathbf{M}_\pi[(\mathbf{v}^{(n)}{}_j - a_j)^2] \to 0$ as n tends to infinity. To prove this it is sufficient to prove that, for every i, $\mathbf{M}_i[(\mathbf{v}^{(n)}{}_j - a_j)^2] \to 0$.

$$\begin{aligned} \mathbf{M}_i[(\mathbf{v}^{(n)}{}_j - a_j)^2] &= \mathbf{M}_i\left[\left(\sum_{k=1}^{n}\left(\mathbf{u}^{(k)}{}_j/n\right) - a_j\right)^2\right] \\ &= \frac{1}{n^2}\mathbf{M}_i\left[\left(\sum_{k=1}^{n}\left(\mathbf{u}^{(k)}{}_j - a_j\right)\right)^2\right]. \end{aligned}$$

Let $m_{k,l} = \mathbf{M}_i[(\mathbf{u}^{(k)}{}_j - a_j)(\mathbf{u}^{(l)}{}_j - a_j)]$. Then we must prove that

$$\frac{1}{n^2} \cdot \sum_{l=1}^{n} \sum_{k=1}^{n} m_{k,l} \to 0 \tag{1}$$

as n tends to infinity. Multiplying out the expression for $m_{k,l}$ we have

$$m_{k,l} = \mathbf{M}_i[\mathbf{u}^{(k)}{}_j\mathbf{u}^{(l)}{}_j] - a_j\mathbf{M}_i[\mathbf{u}^{(k)}{}_j] - a_j\mathbf{M}_i[\mathbf{u}^{(l)}{}_j] + a^2{}_j.$$

Let $m = \min(k, l)$ and $d = |k-l|$. Then

$$m_{k,l} = p^{(m)}{}_{ij}p^{(d)}{}_{jj} - a_jp^{(k)}{}_{ij} - a_jp^{(l)}{}_{ij} + a^2{}_j.$$

Using Corollary 4.1.5, we have

$$m_{k,l} = a_j(e^{(m)}{}_{ij} + e^{(d)}{}_{jj} - e^{(k)}{}_{ij} - e^{(l)}{}_{ij}) + e^{(m)}{}_{ij}e^{(d)}{}_{jj}$$

where $|e^{(n)}{}_{ij}| \leqslant br^n$ with $0 < r < 1$. Hence for a suitably chosen constant c,

$$|m_{k,l}| \leqslant c(r^m + r^d + r^k + r^l). \tag{2}$$

Each value of m, d, k, and l occurs $\leqslant 2n$ times in the sum in (1). Hence, using (2), we have

$$\frac{1}{n^2}\cdot\sum_{l=1}^{n}\sum_{k=1}^{n}|m_{k,l}| \leqslant \frac{4c}{n^2}\cdot\frac{2n}{1-r} = \frac{8c}{n(1-r)}.$$

The right side of this inequality tends to 0 as n tends to infinity; hence, also the left side, as was to be proved.

Let us apply this theorem to the Land of Oz example. We found in § 4.1.5 that $\alpha = (^2/_5, ^1/_5, ^2/_5)$. Thus we can now say that for a large number of days we can expect about $^2/_5$ of the days to be rainy, $^1/_5$ of the days to be nice, and $^2/_5$ of the days to be snowy.

Consider the special case of an independent trials process. Such a process is a Markov chain with transition matrix having all rows the same vector α and with initial probability vector chosen to be α. The law of large numbers for an independent trials process is thus a special case of the theorem just proved. The proof for this case is very much simpler. In fact, in this case $\mathbf{M}_\alpha[\mathbf{u}^{(n)}{}_j] = a_j$ for all n; hence also $\mathbf{M}_\alpha[\mathbf{v}^{(n)}{}_j] = a_j$. Also $m_{k,l} = 0$ for $k \neq l$ and $m_{k,k} = \sigma^2$, a constant for all k. Hence

$$\mathbf{M}_\alpha[(\mathbf{v}^{(n)}{}_j - a_j)^2] = \mathbf{Var}_\alpha[\mathbf{v}^{(n)}{}_j] = \frac{\sigma^2}{n}.$$

This tends to 0 as n tends to infinity.

Another special case of interest is a general Markov chain process which is started by an initial probability vector $\pi = \alpha$. In this case also $\mathbf{M}_\alpha[\mathbf{u}^{(n)}{}_j] = \mathbf{M}_\alpha[\mathbf{v}^{(n)}{}_j] = a_j$ for all n. Hence

$$\mathbf{M}_\alpha[(\mathbf{v}^{(n)}{}_j - a_j)^2] = \mathbf{Var}_\alpha[\mathbf{v}^{(n)}{}_j].$$

However, it is not possible, in this case, to give a simple expression for this variance as a function of n, as was possible in the independent case.

We shall consider this variance in § 4.6, where we shall give an asymptotic expression for it.

§ **4.3 The fundamental matrix for regular chains.** In Chapter III we found that, for an absorbing chain, the matrix $(I-Q)^{-1}$ played a fundamental role. (Q was the matrix obtained by truncating the transition matrix to include only the non-absorbing states.) We shall see that there is a corresponding fundamental matrix for regular chains.

4.3.1 THEOREM. *Let P be the transition matrix for a regular Markov chain. Let A be the limiting matrix. Then $Z=(I-(P-A))^{-1}$ exists and*

$$Z = I + \sum_{n=1}^{\infty} (P^n - A).$$

PROOF. We shall prove that $(P-A)^n = P^n - A$. Since $P^n - A \to 0$, our theorem will then follow from the matrix theorem proved in § 1.11.1. We have $A^2 = \xi\alpha\xi\alpha = \xi\alpha = A$, hence $A^k = A$, and

$$\begin{aligned}(P-A)^n &= \sum_{i=0}^{n} \binom{n}{i} (-1)^{n-i} P^i A^{n-i} \\ &= P^n + \sum_{i=0}^{n-1} \binom{n}{i} (-1)^{n-i} A \\ &= P^n - A.\end{aligned}$$

4.3.2 DEFINITION. *Let P be a regular transition matrix. The matrix $Z=(I-(P-A))^{-1}$ is called the* fundamental matrix *for the Markov chain determined by P.*

We shall see that the matrix Z is the basic quantity used to compute most of the interesting descriptive quantities for the behavior of a regular Markov chain. We shall first establish certain important properties of the matrix which will be useful in later work.

4.3.3 THEOREM. *Let Z be the fundamental matrix for a regular Markov chain with transition matrix P, limiting vector α, and limiting matrix A. Then*

(a) $PZ = ZP$

(b) $Z\xi = \xi$

(c) $\alpha Z = \alpha$

(d) $I - Z = A - PZ.$

PROOF. Part (a) follows from the infinite series representation for Z and the fact that P commutes with each term in this infinite series.

Part (b) states that Z has row-sums 1. This again follows from the infinite series representation for Z since the first matrix I has row-sums 1 and each of the matrices $P^n - A$ have row-sums 0. Part (c) follows from the infinite series representation for Z, since $\alpha I = \alpha$ and $\alpha(P^n - A) = 0$. To prove (d) we multiply $Z = I + \sum_{n=1}^{\infty} (P^n - A)$ by $I - P$ obtaining

$$\begin{aligned}(I-P)Z &= (I-P)+(P-A)\\ &= I-A.\end{aligned}$$

We shall use the Land of Oz example as our standard example of the applications of the Z matrix. For this example P and A are

$$P = \begin{matrix} & \begin{matrix} \mathbf{R} & \mathbf{N} & \mathbf{S} \end{matrix} \\ \begin{matrix} \mathbf{R} \\ \mathbf{N} \\ \mathbf{S} \end{matrix} & \begin{pmatrix} 1/2 & 1/4 & 1/4 \\ 1/2 & 0 & 1/2 \\ 1/4 & 1/4 & 1/2 \end{pmatrix} \end{matrix} \qquad A = \begin{matrix} \begin{matrix} \mathbf{R} & \mathbf{N} & \mathbf{S} \end{matrix} & \\ \begin{pmatrix} 2/5 & 1/5 & 2/5 \\ 2/5 & 1/5 & 2/5 \\ 2/5 & 1/5 & 2/5 \end{pmatrix} & \begin{matrix} \mathbf{R} \\ \mathbf{N}. \\ \mathbf{S} \end{matrix} \end{matrix}$$

To find the matrix Z we must find the inverse of the matrix

$$I - P + A = \begin{pmatrix} .9 & -.05 & .15 \\ -.1 & 1.2 & -.1 \\ .15 & -.05 & .9 \end{pmatrix}.$$

Doing this we obtain

$$Z = 1/75 \begin{matrix} \begin{matrix} \mathbf{R} & \mathbf{N} & \mathbf{S} \end{matrix} & \\ \begin{pmatrix} 86 & 3 & -14 \\ 6 & 63 & 6 \\ -14 & 3 & 86 \end{pmatrix} & \begin{matrix} \mathbf{R} \\ \mathbf{N}. \\ \mathbf{S} \end{matrix} \end{matrix}$$

While the fundamental matrix Z has several properties in common with a transition matrix, we see from this example that it does not necessarily have non-negative entries.

An example where the Z matrix turns out to be a very simple matrix is the case of an independent trials process. In this case $P = A$ so that $Z = (I - (P - A))^{-1} = I$. Thus for an independent trials process the Z matrix is the identity matrix.

Let $\bar{\mathrm{y}}^{(n)}{}_j$ be the number of times that the process is in state s_j in the first n stages, i.e. the initial position plus $n-1$ stages.

4.3.4 THEOREM. *For any regular Markov chain, and any initial vector* π,

$$\{\mathbf{M}_\pi[\bar{\mathbf{y}}^{(n)}{}_j]\} - n\alpha \to \pi(Z-A) = \pi Z - \alpha.$$

PROOF. For any i,

$$\mathbf{M}_i[\bar{\mathbf{y}}^{(n)}{}_j] = \sum_{k=0}^{n-1} \mathbf{M}_i[\mathbf{u}^{(k)}{}_j] = \sum_{k=0}^{n-1} p^{(k)}{}_{ij}.$$

Thus

$$\{\mathbf{M}_i[\bar{\mathbf{y}}^{(n)}{}_j] - na_j\} = \sum_{k=0}^{n-1} (P^k - A) \to Z - A.$$

Therefore

$$\pi\{\mathbf{M}_i[\bar{\mathbf{y}}^{(n)}{}_j] - na_j\} \to \pi(Z-A) = \pi Z - \alpha.$$

An immediate consequence of this theorem is the following:

4.3.5 COROLLARY. *For any two initial distributions* π *and* π'

$$\{\mathbf{M}_\pi[\bar{\mathbf{y}}^{(n)}{}_j] - \mathbf{M}_{\pi'}[\bar{\mathbf{y}}^{(n)}{}_j]\} \to (\pi - \pi')Z.$$

If we choose a particular starting state, say i, then Theorem 4.3.4 states that

$$\mathbf{M}_i[\bar{\mathbf{y}}^{(n)}{}_j] - na_j \to (z_{ij} - a_j).$$

Thus we see that for large n the mean number of times in state s_j, starting at state s_i, differs from na_j by approximately $z_{ij} - a_j$ We recall that by Theorem 4.2.1 the mean of the fraction of times in state s_j approaches a_j independent of the starting state. Thus the entries of $(Z-A)$ give us an interesting quantity for regular chains for which the initial state does have an influence. We can compare two starting states, since by Corollary 4.3.5

$$\mathbf{M}_i[\bar{\mathbf{y}}^{(n)}{}_j] - \mathbf{M}_k[\bar{\mathbf{y}}^{(n)}{}_j] \to z_{ij} - z_{kj}.$$

Another interesting corollary to Theorem 4.3.4 is the following.

4.3.6 COROLLARY. *Let* $c = \sum z_{jj}$. *Then*

$$\sum_j (\mathbf{M}_j[\bar{\mathbf{y}}^{(n)}{}_j] - \mathbf{M}_\pi[\bar{\mathbf{y}}^{(n)}{}_j]) \to c - 1$$

as n *approaches infinity, independent of* π.

PROOF. By Corollary 4.3.5

$$\mathbf{M}_j[\bar{\mathbf{y}}^{(n)}{}_j] - \mathbf{M}_\pi[\bar{\mathbf{y}}^{(n)}{}_j] \to z_{jj} - (\pi Z)_j.$$

Therefore the sum approaches

$$\sum z_{jj} - \pi Z \xi = c - 1.$$

This corollary has the following interpretation. For any π, $\mathbf{M}_j[\bar{\mathbf{y}}^{(n)}{}_j] \geqslant \mathbf{M}_\pi[\bar{\mathbf{y}}^{(n)}{}_j]$. Hence $\mathbf{M}_j[\bar{\mathbf{y}}^{(n)}{}_j]$ gives the largest possible mean number of times in $\mathbf{s}_j$. The corollary states that the deviations from this maximum, summed over all states, approach a limit which is independent of the choice of π.

§ **4.4 First passage times.** In this section we shall study the length of time to go from a state $\mathbf{s}_i$ to a state $\mathbf{s}_j$ for the first time. We shall see that the mean of this time is easily obtained from the fundamental matrix.

4.4.1 DEFINITION. *For a regular Markov chain, the* first passage time $\mathbf{f}_k$ *is a function whose value is the number of steps before entering* $\mathbf{s}_k$ *for the first time after the initial position.*

4.4.2 THEOREM. *For any* i, $\mathbf{M}_i[\mathbf{f}_k]$ *is finite.*

PROOF. Assume first that $i \neq k$. Form a new Markov chain by making state $\mathbf{s}_k$ into an absorbing state. The resulting Markov chain is an absorbing Markov chain with a single absorbing state, $\mathbf{s}_k$. The mean time to go from $\mathbf{s}_i$ to $\mathbf{s}_j$ in the given chain is the same as the mean time before absorption in the new chain. The mean time before absorption is finite by Theorem 3.2.4.

If $i = k$, then

$$\mathbf{M}_i[\mathbf{f}_i] = p_{ii} + \sum_{k \neq i} p_{ik}\mathbf{M}_k[\mathbf{f}_i]$$

which is finite by the first part of the proof.

4.4.3 DEFINITION. *The* mean first passage matrix, *denoted by* M, *is the matrix with entries* $m_{ij} = \mathbf{M}_i[\mathbf{f}_j]$.[†]

It then follows from 1.8.9 that, for an initial vector π, the mean first passage times are the components of the vector πM.

4.4.4 THEOREM. *The matrix* M *satisfies the equation*

$$M = P(M - M_{dg}) + E. \tag{1}$$

PROOF. We calculate $\mathbf{M}_i[\mathbf{f}_j]$ by taking the mean of the conditional means, given the outcome of the first experiment. This gives

$$\begin{aligned}\mathbf{M}_i[\mathbf{f}_j] &= \sum_{k \neq j} p_{ik}(\mathbf{M}_k[\mathbf{f}_j] + 1) + p_{ij} \\ &= \sum_{k \neq j} p_{ik}\mathbf{M}_k[\mathbf{f}_j] + 1 \\ &= \sum_{k} p_{ik}\mathbf{M}_k[\mathbf{f}_j] - p_{ij}\mathbf{M}_j[\mathbf{f}_j] + 1.\end{aligned}$$

That is,

$$m_{ij} = \sum_k p_{ik}m_{kj} - p_{ij}m_{jj} + 1.$$

This proves the theorem.

† This matrix is denoted by $\overline{M}$ in later works by the authors.

4.4.5 THEOREM. *Let $\alpha = \{a_1, a_2, \ldots, a_r\}$ be the limiting probability vector for P. Then $m_{ii} = 1/a_i$.*

PROOF. Multiplying equation (1) above by α we have

$$\begin{aligned} \alpha M &= \alpha P(M - M_{\mathrm{dg}}) + \alpha E \\ &= \alpha(M - M_{\mathrm{dg}}) + \alpha E. \end{aligned}$$

Therefore

$$\alpha M_{\mathrm{dg}} = \alpha E = \eta.$$

This states that $\alpha_i m_{ii} = 1$ for every i, or $m_{ii} = 1/a_i$.

4.4.6 THEOREM. *Equation* (1) *of Theorem* 4.4.4 *has a unique solution.*

PROOF. Let M and M' be two solutions for (1). Then from the proof of Theorem 4.4.5 we have $\alpha M_{\mathrm{dg}} = \alpha M'_{\mathrm{dg}} = \eta$. Hence $M_{\mathrm{dg}} = M'_{\mathrm{dg}}$. This gives

$$M - M' = P(M - M').$$

But this means that each column of $M - M'$ is a fixed column vector for P. Hence by Theorem 4.1.7 each column is a constant vector. Since $M - M'$ has 0's on the diagonal, these vectors must all be 0 vectors. Hence $M = M'$.

4.4.7 THEOREM. *The mean first passage matrix M is given by*

$$M = (I - Z + EZ_{\mathrm{dg}})D \tag{2}$$

where D is the diagonal matrix with diagonal elements $d_{ii} = 1/a_i$.

PROOF. By Theorems 4.4.4 and 4.4.6 we need only show that M as defined by (2) satisfies equation (1) above.

Let

$$M = (I - Z + EZ_{\mathrm{dg}})D.$$

Then

$$M - D = (-Z + EZ_{\mathrm{dg}})D$$

and

$$\begin{aligned} P(M - D) &= (-PZ + EZ_{\mathrm{dg}})D \\ &= M + (-I + Z - PZ)D. \end{aligned}$$

By Theorem 4.3.3(d) this is

$$\begin{aligned} P(M - D) &= M - AD \\ &= M - E. \end{aligned}$$

By (2), $D = M_{\mathrm{dg}}$. Hence $M = P(M - M_{\mathrm{dg}}) + E$.

4.4.8 THEOREM. *Let P be the transition matrix for an independent trials process. Then $M = \{1/p_{ij}\}$.*

PROOF. From Theorem 4.4.7 and the fact that $Z = I$ for an independent trials process, we have $M = ED$. For an independent trials process the limit matrix $A = P$. Hence $p_{ij} = a_j$ and $1/a_j = 1/p_{ij}$. Thus $M = ED = \{1/p_{ij}\}$.

We now illustrate the calculation of the mean first passage matrix for the Land of Oz example. We have found α to be $(^2/_5, {}^1/_5, {}^2/_5)$. Hence the matrix D is

$$D = \begin{pmatrix} ^5/_2 & 0 & 0 \\ 0 & 5 & 0 \\ 0 & 0 & ^5/_2 \end{pmatrix}.$$

The Z matrix was found in § 4.3. From this, using Theorem 4.4.7 we obtain M by $M = (I - Z + EZ_{dg})D$. Carrying out this calculation we obtain

$$M = \begin{array}{c} \\ \mathbf{R} \\ \mathbf{N} \\ \mathbf{S} \end{array} \begin{array}{c} \begin{array}{ccc} \mathbf{R} & \mathbf{N} & \mathbf{S} \end{array} \\ \begin{pmatrix} ^5/_2 & 4 & ^{10}/_3 \\ ^8/_3 & 5 & ^8/_3 \\ ^{10}/_3 & 4 & ^5/_2 \end{pmatrix} \end{array}.$$

Thus, for example, if it is raining in the Land of Oz today the mean number of days before a nice day is 4. The mean number of days before another rainy day is $^5/_2$; before a snowy day $^{10}/_3$.

We shall next prove a theorem which connects the diagonal elements of Z with the mean time to reach s_j for the initial probability vector $\pi = \alpha$. We have seen previously that the mean number of times in state s_j is particularly simple in this case. This choice of initial vector is of special importance for the following reason. Assume that a regular Markov chain has gone through a large number of steps before it is observed. Then Theorem 4.1.6 suggests that a natural choice for the new initial vector π is α. The probabilities for any later time are then also given by α. In this case we say that the process is observed *in equilibrium.*

4.4.9 THEOREM. *For a regular Markov chain*

$$\alpha M = \{\mathrm{M}_\alpha[\mathrm{f}_j]\} = \eta Z_{dg} D = \{z_{jj}/a_j\}.$$

PROOF. Multiplying (2) by α we have

$$\begin{aligned} \alpha M &= \alpha(I - Z + EZ_{dg})D \\ &= (\alpha - \alpha + \eta Z_{dg})D \\ \alpha M &= \eta Z_{dg} D. \end{aligned}$$

4.4.10 THEOREM. *Let* $c = \sum_i z_{ii}$. *Then* $M\alpha^T = c\xi$.

PROOF.

$$\begin{aligned} M\alpha^T &= (I - Z + EZ_{dg})D\alpha^T \\ &= (I - Z + EZ_{dg})\xi \\ &= \xi(\eta Z_{dg}\xi) = c\xi. \end{aligned}$$

In § 4.3 we compared the mean number of times $\bar{\mathbf{y}}^{(n)}{}_j$ in a state s_j under the assumption of two different starting distributions. We can make the same comparison for the function $\mathbf{f}_j$.

4.4.11 THEOREM. *For any two initial probability vectors* π *and* π'

$$\{\mathbf{M}_\pi[\mathbf{f}_j] - \mathbf{M}_{\pi'}[\mathbf{f}_j]\} = (\pi - \pi')(I - Z)D.$$

PROOF.

$$\begin{aligned} \{\mathbf{M}_\pi[\mathbf{f}_j] - \mathbf{M}_{\pi'}[\mathbf{f}_j]\} &= \pi M - \pi' M \\ &= (\pi - \pi')(I - Z + EZ_{dg})D. \\ &= (\pi - \pi')(I - Z)D. \end{aligned}$$

In the Land of Oz example we see that

$$\mathbf{M}_N[\mathbf{f}_S] - \mathbf{M}_R[\mathbf{f}_S] = {}^8/_3 - {}^{10}/_3 = -{}^2/_3.$$

Thus the mean time to the first snowy day is shorter starting with a nice day than it is starting with a rainy day.

We will conclude this section by showing that the Markov chain is completely determined by the numbers m_{ij}, for $i \neq j$. We shall use these numbers as the non-zero entries of the matrix $\bar{M} = M - D$. This matrix has $n(n-1)$ non-zero entries, which suffices to determine the chain. When we give the chain in terms of P, we specify n^2 entries, but these satisfy n relations, since P must have row-sums 1. But there is no natural way of specifying just $n(n-1)$ entries of P, while $\bar{M}$ is a natural way of giving this minimum information.

4.4.12 THEOREM. *For any regular Markov chain*

(a) *The matrix* $\bar{M}$ *has an inverse*

(b) $\alpha = (c-1)\,(\bar{M}^{-1}\xi)^T$

(c) $P = I + (D - E)\bar{M}^{-1}$.

PROOF. From equation (1) in § 4.4.4 we have

$$\bar{M} + D = P\bar{M} + E;$$

hence

$$(P - I)\bar{M} = D - E. \tag{3}$$

If $\bar{M}$ has no inverse, then there is a non-zero column vector γ such that $\bar{M}\gamma = 0$. Hence from (3)

$$(D-E)\gamma = (P-I)\bar{M}\gamma = 0$$
$$D\gamma = E\gamma$$
$$\gamma = D^{-1}E\gamma = D^{-1}\xi\eta\gamma = (\eta\gamma)\alpha^T,$$

where $l = \eta\gamma$ is a number. And since $\gamma \neq 0$, $l \neq 0$.

$$\alpha^T = (^1/l)\gamma$$
$$\bar{M}\alpha^T = (^1/l)\bar{M}\gamma = 0.$$

But, clearly, $\bar{M}\alpha^T > 0$, and we have a contradiction. Therefore, $\bar{M}$ has an inverse. Formula (c) is then an immediate consequence of (3). To prove (b) we make use of § 4.4.9 and the fact that $D\alpha^T = \xi$.

$$(\bar{M}+D)\alpha^T = c\xi$$
$$\bar{M}\alpha^T = (c-1)\xi$$
$$\alpha^T = (c-1)\bar{M}^{-1}\xi$$
$$\alpha = (c-1)(\bar{M}^{-1}\xi)^T.$$

We can now find α from formula (b), and the condition that $\alpha\xi = 1$. This determines D, and then formula (c) will yield P. Thus the chain is determined by $\bar{M}$.

§ **4.5 Variance of the first passage time.** In the previous section we found that the Z matrix enabled us to find the mean first passage time from $\mathbf{s}_i$ to $\mathbf{s}_j$. In this section we shall show that the Z matrix also provides us with the variance of the first passage time.

We recall that $\mathbf{f}_j$ is the function whose value gives the number of steps required to reach $\mathbf{s}_j$ for the first time after the initial step. We have found $\mathbf{M}_i[\mathbf{f}_j]$. Hence to find $\mathbf{Var}_i[\mathbf{f}_j]$ it is only necessary to find $\mathbf{M}_i[\mathbf{f}^2{}_j]$ and use the fact that $\mathbf{Var}_i[\mathbf{f}_j] = \mathbf{M}_i[\mathbf{f}^2{}_j] - \mathbf{M}_i[\mathbf{f}_j]^2$. We denote by W the matrix $W = \{\mathbf{M}_i[\mathbf{f}^2{}_j]\}$.

4.5.1 THEOREM. *The matrix W satisfies the equation*

$$W = P[W - W_{dg}] - 2P[Z - EZ_{dg}]D + E. \tag{1}$$

PROOF. Taking conditional means we have

$$\mathbf{M}_i[\mathbf{f}^2{}_j] = \sum_{k \neq j} p_{ik}\mathbf{M}_k[(\mathbf{f}_j + 1)^2] + p_{ij}$$
$$= \sum_{k \neq j} p_{ik}\mathbf{M}_k[\mathbf{f}^2{}_j] + 2\sum_{k \neq j} p_{ik}\mathbf{M}_k[\mathbf{f}_j] + 1.$$

Or

$$W = P[W - W_{dg}] + 2P[M - M_{dg}] + E. \tag{2}$$

From Theorem 4.4.7 we have $M - M_{dg} = (-Z + EZ_{dg})D$. Putting this in (2) we have (1).

4.5.2 THEOREM. *The values for* $\mathbf{M}_i[\mathbf{f}^2{}_i]$ *are given by*

$$W_{dg} = D(2Z_{dg}D - I). \tag{3}$$

PROOF. Multiplying equation (1) through by α and using the fact that $\alpha P = \alpha$, we have

$$\alpha W = \alpha[W - W_{dg}] - 2\alpha[Z - EZ_{dg}]D + \eta; \tag{4}$$

or, since $\alpha Z = \alpha$, and $\alpha D = \alpha E = \eta$,

$$\alpha W_{dg} = -\eta + 2\eta Z_{dg}D.$$

This gives

$$\alpha_i w_{ii} = -1 + 2z_{ii}/a_i$$

or

$$w_{ii} = -\frac{1}{a_i} + \frac{2z_{ii}}{a^2{}_i}.$$

Written in matrix form, this is (3).

4.5.3 THEOREM. *The unique solution to* (1) *is*

$$W = M(2Z_{dg}D - I) + 2(ZM - E(ZM)_{dg}).$$

PROOF. The uniqueness proof is the same as that given for the matrix M in Theorem 4.4.6. It is then only a matter of verifying that the expression given for W satisfies (1). We omit the details of this.

From the matrix W it is an easy matter to find the $\{\mathbf{Var}_i[\mathbf{f}_j]\}$. We denote by $M_2 = \{\mathbf{Var}_i[\mathbf{f}_j]\}$. Then $M_2 = W - M_{sq}$.

Let us find these variances for the Land of Oz example. We have previously found M, D, and Z for this example, so that to find W, the only new matrix we need is ZM. This is

$$Z \cdot M = {}^1/_{75}\begin{pmatrix} 176^1/_3 & 303 & 259^2/_3 \\ 203 & 363 & 203 \\ 259^2/_3 & 303 & 176^1/_3 \end{pmatrix}.$$

From the formula $W = M(2Z_{dg}D - I) + 2(ZM - E(ZM)_{dg})$ we find

$$W = \begin{pmatrix} {}^{71}/_6 & 26 & {}^{54}/_3 \\ {}^{40}/_3 & 37 & {}^{40}/_3 \\ {}^{54}/_3 & 28 & {}^{71}/_6 \end{pmatrix},$$

and subtracting M_{sq} from this we obtain

$$M_2 = \begin{pmatrix} ^{67}/_{12} & 12 & ^{62}/_9 \\ ^{56}/_9 & 12 & ^{56}/_9 \\ ^{62}/_9 & 12 & ^{67}/_{12} \end{pmatrix}.$$

We observe that the variance $\mathbf{Var}_i[\mathbf{f}_j]$ in this example depends very little on the choice of the starting state $\mathbf{s}_i$. The first passage times for a regular chain are quite similar to absorption times for an absorbing Markov chain. They both have variances which are in general large compared to their means.

The formulas for W and for M_2 are very much simplified for the case of an independent trials process.

4.5.4 THEOREM. *For an independent trials process*

$$W = ED(2D-I) = \{(1/p_{ij})(2/p_{ij}-1)\}$$

and

$$M_2 = E(D^2-D) = \{(1/p_{ij})^2-1/p_{ij}\}.$$

PROOF. We recall that for an independent trials process Z is the identity matrix and $M=ED$. Thus, using Theorem 4.5.3,

$$\begin{aligned} W &= ED(2D-I)+2(ED-ED) \\ &= ED(2D-I). \end{aligned}$$

From this we obtain

$$\begin{aligned} M_2 &= W-M_{sq} \\ &= 2ED^2-ED-(ED)_{sq} \\ &= E(D^2-D). \end{aligned}$$

The alternative expressions for W and M_2 given in the statement of the theorem follow from the fact that $p_{ij}=p_{jj}$ for all i.

§ **4.6 Limiting covariance.** Let $\mathbf{f}$ and $\mathbf{g}$ be two functions defined on the states of a regular chain. Let $\mathbf{f}(\mathbf{s}_i)=f_i$ and $\mathbf{g}(\mathbf{s}_i)=g_i$. Let $\mathbf{f}^{(n)}$ and $\mathbf{g}^{(n)}$ be the values of these functions on the n-th step. We are interested in finding

$$\lim_{n\to\infty}\frac{1}{n}\mathbf{Cov}_\pi\left[\sum_{k=1}^{n}\mathbf{f}^{(k)}, \sum_{k=1}^{n}\mathbf{g}^{(k)}\right].$$

It can be shown that this limit exists and is independent of π.

4.6.1 THEOREM.

$$\lim_{n\to\infty}\frac{1}{n}\mathbf{Cov}_\pi\left[\sum_{k=1}^{n}\mathbf{f}^{(k)}, \sum_{l=1}^{n}\mathbf{g}^{(k)}\right] = \sum_{i,j=1}^{r} f_i c_{ij} g_j$$

where

$$c_{ij} = a_i z_{ij} + a_j z_{ji} - a_i d_{ij} - a_i a_j.$$

PROOF. We shall assume the result that the limit is independent of π and prove the theorem for the case $\pi = \alpha$.

$$\mathbf{M}_\alpha\left[\sum_{k=1}^{n} \mathbf{f}^{(k)}\right] = n \sum_{i=1}^{r} a_i f_i$$

and

$$\mathbf{M}_\alpha\left[\sum_{k=1}^{n} \mathbf{g}^{(k)}\right] = n \sum_{j=1}^{r} a_j g_j.$$

Hence,

$$\frac{1}{n} \mathbf{Cov}_\alpha\left[\sum_{k=1}^{n} \mathbf{f}^{(k)}, \sum_{l=1}^{n} \mathbf{g}^{(l)}\right]$$

$$= \frac{1}{n} \mathbf{M}_\alpha\left[\left(\sum_{k=1}^{n} \mathbf{f}^{(k)} - n \sum_{i=1}^{r} a_i f_i\right)\left(\sum_{l=1}^{n} \mathbf{g}^{(l)} - n \sum_{j=1}^{r} a_j g_j\right)\right]$$

$$= \frac{1}{n} \mathbf{M}_\alpha\left[\sum_{k=1}^{n} \sum_{l=1}^{n} \mathbf{f}^{(k)}\mathbf{g}^{(l)} - n \sum_{k=1}^{n} \mathbf{f}^{(k)} \sum_{j=1}^{r} a_j g_j\right.$$

$$\left. - n \sum_{l=1}^{n} \mathbf{g}^{(l)} \sum_{i=1}^{r} a_i f_i + n^2 \sum_{i,j=1}^{r} a_i a_j f_i g_j\right]$$

$$= \frac{1}{n} \sum_{k,l=1}^{n} \sum_{i,j=1}^{r} (\mathbf{Pr}_\alpha[\mathbf{u}^{(k)}{}_i = 1 \wedge \mathbf{u}^{(l)}{}_j = 1] f_i g_j - \mathbf{Pr}_\alpha[\mathbf{u}^{(k)}{}_i = 1] f_i a_j g_j$$

$$- \mathbf{Pr}_\alpha[\mathbf{u}^{(l)}{}_j = 1] g_j a_i f_i + a_i a_j f_i g_j)$$

$$= \frac{1}{n} \sum_{k,l=1}^{n} \sum_{i,j=1}^{r} (\mathbf{Pr}_\alpha[\mathbf{u}^{(k)}{}_i = 1 \wedge \mathbf{u}^{(l)}{}_j = 1] f_i g_j - a_i a_j f_i g_j). \tag{1}$$

Now

$$\mathbf{Pr}_\alpha[\mathbf{u}^{(k)}{}_i = 1 \wedge \mathbf{u}^{(l)}{}_j = 1] = \begin{cases} a_i p^{(l-k)}{}_{ij} & \text{if } k < l \\ a_j p^{(k-l)}{}_{ji} & \text{if } k > l \\ a_i d_{ij} & \text{if } k = l. \end{cases} \tag{2}$$

Hence we have from (1) and (2)

$$\frac{1}{n} \mathbf{Cov}_\alpha\left[\sum_{k=1}^{n} \mathbf{f}^{(k)}, \sum_{l=1}^{n} \mathbf{g}^{(l)}\right] = \frac{1}{n} \sum_{i,j=1}^{r} a_i f_i g_j \sum_{k<l} (p^{(l-k)}{}_{ij} - a_j)$$

$$+ \frac{1}{n} \sum_{i,j=1}^{r} a_j f_i g_j \sum_{k>l} (p^{(k-l)}{}_{ji} - a_i)$$

$$+ \sum_{i,j=1}^{r} (a_i d_{ij} - a_i a_j) f_i g_j.$$

Collecting terms with the same $d = |l-k|$, we have

$$\frac{1}{n}\,\mathbf{Cov}_\alpha\left[\sum_{k=1}^{n}\mathbf{f}^{(k)},\ \sum_{l=1}^{n}\mathbf{g}^{(l)}\right] = \sum_{i,j=1}^{r} a_i f_i g_j \sum_{d=1}^{n-1}\frac{n-d}{n}\,(p^{(d)}{}_{ij}-a_j)$$
$$+\sum_{i,j=1}^{r} a_j f_i g_j \sum_{d=1}^{n-1}\frac{n-d}{n}\,(p^{(d)}{}_{ji}-a_i) \qquad (3)$$
$$+\sum_{i,j=1}^{r}(a_i d_{ij}-a_i a_j) f_i g_j.$$

Since $Z = \sum_{d=0}^{\infty}(P-A)^d$ converges, it is Cesaro-summable (see § 1.10), that is

$$Z = \lim_{n\to\infty}\sum_{d=0}^{n-1}\frac{n-d}{n}\,(P-A)^d.$$

Hence

$$Z-I = \lim_{n\to\infty}\sum_{d=1}^{n-1}\frac{n-d}{n}\,(P^d-A),$$

$$z_{ij}-d_{ij} = \lim_{n\to\infty}\sum_{d=1}^{n-1}\frac{n-d}{n}\,(p^{(d)}{}_{ij}-a_j).$$

Then from (3)

$$\lim_{n\to\infty}\frac{1}{n}\,\mathbf{Cov}_\alpha\left[\sum_{k=1}^{n}\mathbf{f}^{(k)},\ \sum_{l=1}^{n}\mathbf{g}^{(l)}\right]$$
$$= \sum_{i,j=1}^{r}\left[a_i f_i g_j (z_{ij}-d_{ij}) + a_j f_i g_j (z_{ji}-d_{ij}) + (a_i d_{ij} - a_i a_j) f_i g_j\right]$$
$$= \sum_{i,j=1}^{r} f_i (a_i z_{ij} + a_j z_{ji} - a_i d_{ij} - a_i a_j) g_j.$$

This completes the proof.

If **f** and **g** are the same function, then the above theorem gives us

4.6.2 COROLLARY.

$$\lim_{n\to\infty}\frac{1}{n}\,\mathbf{Var}_\pi\left[\sum_{k=1}^{n}\mathbf{f}^{(k)}\right] = \sum_{i,j=1}^{r} f_i c_{ij} f_j.$$

We shall need a slight extension of this last result. Suppose that **f** is not simply a function of the state, but $\mathbf{f} = 1$ with probability f_i on state s_i and 0 with probability $1-f_i$. We may think of **f** as determined as follows: We carry out the Markov chain, and if the process is in s_i on a given step we flip a biased coin (probability f_i for heads) to determine whether **f** is 1 or 0. Again, $\mathbf{f}^{(n)}$ is the value of the function on the n-th step. Then the above argument for the limiting variance applies except that in (1) a slight change must be made when $k=l$. Here, the term f^2_i should be simply f_i, and we have a correction term

$$\sum_{i=1}^{r} a_{ii} f_i(1-f_i).$$

Thus we have proved

4.6.3 THEOREM. *If* **f** *is a function that takes on the value* 1 *with probability* f_i *in* s_i, *and is* 0 *otherwise, then*

$$\lim_{n\to\infty} \frac{1}{n} \mathbf{Var}_\pi\left[\sum_{k=1}^{n} \mathbf{f}^{(k)}\right] = \sum_{i,j=1}^{r} f_i c_{ij} f_j + \sum_{i=1}^{r} a_{ii} f_i(1-f_i).$$

We can also extend the result to two such functions. If these take on their values independently of each other, then the proof of 4.6.1 applies exactly. This proves

4.6.4 THEOREM. *If* **f** *and* **g** *are functions that take on the value* 1 *in* s_i *with probabilities* f_i *and* g_i, *respectively, independently of each other, and if the functions are* 0 *otherwise, then*

$$\lim_{n\to\infty} \frac{1}{n} \mathbf{Cov}_\pi\left[\sum_{k=1}^{n} \mathbf{f}^{(k)}, \sum_{l=1}^{n} \mathbf{g}^{(l)}\right] = \sum_{i,j=1}^{r} f_i c_{ij} g_j.$$

One application of the covariance is to obtain correlation coefficients. Let **f** and **g** be as in Theorem 4.6.1. Then

4.6.5 DEFINITION.

$$\mathbf{Corr}_\pi\left[\sum_{k=1}^{n} \mathbf{f}^{(k)}, \sum_{l=1}^{n} \mathbf{g}^{(l)}\right] = \frac{\mathbf{Cov}_\pi\left[\sum_{k=1}^{n} \mathbf{f}^{(k)}, \sum_{l=1}^{n} \mathbf{g}^{(l)}\right]}{\sqrt{\mathbf{Var}_\pi\left[\sum_{k=1}^{n} \mathbf{f}^{(k)}\right] \cdot \mathbf{Var}_\pi\left[\sum_{l=1}^{n} \mathbf{g}^{(l)}\right]}}.$$

Dividing numerator and denominator of the right side by n, and using 4.6.1 and 4.6.2, we have

4.6.6 THEOREM.

$$\lim_{n\to\infty} \mathbf{Corr}_\pi\left[\sum_{k=1}^{n} \mathbf{f}^{(k)}, \sum_{l=1}^{n} \mathbf{g}^{(l)}\right] = \frac{\sum_{i,j=1}^{r} f_i c_{ij} g_j}{\sqrt{\sum_{i,j=1}^{r} f_i c_{ij} f_j \cdot \sum_{i,j=1}^{r} g_i c_{ij} g_j}}.$$

Another important application of Theorem 4.6.1 is the following: Let A and B be any two sets of states. Let $\mathbf{y}^{(n)}{}_A$ and $\mathbf{y}^{(n)}{}_B$ be respectively the number of times in set A in the first n steps and the number of times in set B in the first n steps. For these functions we have the following theorem.

4.6.7 THEOREM.

$$\lim_{n\to\infty} \frac{1}{n} \mathbf{Cov}_\pi[\mathbf{y}^{(n)}{}_A, \ \mathbf{y}^{(n)}{}_B] = \sum_{\substack{s_i \text{ in } A \\ s_j \text{ in } B}} c_{ij}.$$

PROOF. Let $\mathbf{f}$ be a function which is 1 on the states of A and 0 on all other states. Let $\mathbf{g}$ be a function which is 1 on states of $\mathbf{B}$ and 0 otherwise.

$$\mathbf{y}^{(n)}{}_A = \sum_{k=1}^{n} \mathbf{f}^{(k)} \quad \text{and} \quad \mathbf{y}^{(n)}{}_B = \sum_{l=1}^{n} \mathbf{g}^{(l)}.$$

Hence the theorem follows from 4.6.1.

From this theorem we see that

4.6.8 COROLLARY.

$$\lim_{n\to\infty} \mathbf{Corr}_\pi[\mathbf{y}^{(n)}{}_A, \ \mathbf{y}^{(n)}{}_B] = \frac{\displaystyle\sum_{\substack{s_i \text{ in } A \\ s_j \text{ in } B}} c_{ij}}{\sqrt{\displaystyle\sum_{\substack{s_i \text{ in } A \\ s_j \text{ in } A}} c_{ij} \cdot \sum_{\substack{s_i \text{ in } B \\ s_j \text{ in } B}} c_{ij}}}.$$

Taking A and B in 4.6.7 to be sets with a single element, we see that c_{ij} represents the limiting covariance for the number of times in states i and j in the first n steps. The values of c_{ii} give the limiting variances for the number of times in state s_i. We are often interested only in these variances, we denote them by the vector β. The limiting correlation for the number of times in s_i and s_j is $\dfrac{c_{ij}}{\sqrt{c_{ii} \cdot c_{jj}}}$. If $i = j$, the correlation is 1.

For an independent trials process $c_{ij} = a_i d_{ij} - a_i a_j$, and hence all the formulas found above simplify. For example, if $i \neq j$, the limiting correlation is

$$\frac{-a_i a_j}{\sqrt{a_i(1-a_i)a_j(1-a_j)}} = -\sqrt{\frac{a_i a_j}{(1-a_i)(1-a_j)}}.$$

The diagonal entries of C, i.e. the limiting variances, have the following important use. Let $\beta = \{b_j\} = \{c_{jj}\}$. Then β is a vector which gives the *limiting variances* for the number of times in each state. These variances appear in the following important theorem (called the Central Limits Theorem for Markov Chains).

4.6.9 THEOREM. *For an ergodic chain, let* $\mathbf{y}^{(n)}{}_j$ *be the number of times in state* $\mathbf{s}_j$ *in the first n steps and let* $\alpha = \{a_j\}$ *and* $\beta = \{b_j\}$ *respectively be the fixed vector and the vector of limiting variances. Then if* $b_j \neq 0$, *for any numbers* $r < s$,

$$\mathbf{Pr}_k\left[r < \frac{\mathbf{y}^{(n)}{}_j - na_j}{\sqrt{nb_j}} < s\right] \to \frac{1}{\sqrt{2\pi}} \int_r^s e^{-x^2/2}\, dx$$

as $n \to \infty$, *for any choice of starting state k.*

The proof of this theorem is beyond the scope of this book and appears only in the more advanced books on probability theory. However, for a discussion of this theorem in the case of independent trials processes see FMS Chapter 3. It is not possible to evaluate the integral in this theorem exactly, but for illustrative purposes we mention that the value for $r = -1$ and $s = 1$ is approximately .681, for $r = -2$ and $s = 2$ it is approximately .954, and for $r = -3$ and $s = 3$ it is .997.

EXAMPLE. Let us consider the Land of Oz example. For this example we have found,

$$\alpha = ({}^2/_5, {}^1/_5, {}^2/_5)$$

and

$$Z = {}^1/_{75} \begin{array}{c} \begin{array}{ccc} \mathbf{R} & \mathbf{N} & \mathbf{S} \end{array} \\ \begin{pmatrix} 86 & 3 & -14 \\ 6 & 63 & 6 \\ -14 & 3 & 86 \end{pmatrix} \end{array}.$$

This is all the information we need to compute the matrix $C = \{c_{ij}\}$. Carrying out this computation we find

$$C = {}^1/_{375} \begin{array}{c} \begin{array}{ccc} \mathbf{R} & \mathbf{N} & \mathbf{S} \end{array} \\ \begin{pmatrix} 134 & -18 & -116 \\ -18 & 36 & -18 \\ -116 & -18 & 134 \end{pmatrix} \end{array} \begin{array}{l} \mathbf{R} \\ \mathbf{N}. \\ \mathbf{S} \end{array}$$

The diagonal entries of C give us the limiting variance $\beta = \{{}^{134}/_{375}, {}^{36}/_{375}, {}^{134}/_{375}\}$ for being in each of the states. Thus the Central Limit Theorem would say, for example, that

$$\frac{\mathbf{y}^{(n)}{}_N - n/5}{\sqrt{({}^{36}/_{375})n}}$$

would for large n have approximately a normal distribution. From this we may estimate that the number of days in 375 days which would be nice would be unlikely (probability about .046) to deviate from 75 by more than 12.

Assume that we are interested only in bad weather or only in good weather. Then we would want to consider the number of times the process is in the set $\mathbf{A}_1 = \{\mathbf{R}, \mathbf{S}\}$ and the number of times it is in the set $\mathbf{A}_2 = \{\mathbf{N}\}$. Let $\hat{c}_{ij}$ be the limiting covariance for the number of times in set $\mathbf{A}_i$ and $\mathbf{A}_j$. Then from 4.6.7 we know that we can obtain $\hat{C} = \{\hat{c}_{ij}\}$ by simply adding elements of C. For example,

$$\hat{c}_{12} = c_{RN} + c_{SN} = -{}^{18}/_{375} - {}^{18}/_{375} = -{}^{36}/_{375} = -{}^{12}/_{125}.$$

$$\hat{C} = \begin{array}{c} \\ \mathbf{A}_1 \\ \mathbf{A}_2 \end{array}\!\!\begin{array}{c} \begin{array}{cc} \mathbf{A}_1 & \mathbf{A}_2 \end{array} \\ \begin{pmatrix} {}^{12}/_{125} & -{}^{12}/_{125} \\ -{}^{12}/_{125} & {}^{12}/_{125} \end{pmatrix} \end{array}.$$

It is easily verified that the row-sums of C must be 0. Since C is symmetric, the column-sums must also be 0. For a 2×2 matrix this tells us that the entries must all have the same absolute value. Thus we would expect $\hat{C}$ to have the special form that we found.

§ 4.7 Comparison of two examples. In this section we shall compare the basic quantities for two regular Markov chains with the same limiting vector. One will be an independent trials example and the other will be a dependent example. The two examples are the random walk Example 3 of Chapter II with $p = {}^1/_2$ (denoted by Example 3a) and Example 7. The transition matrix for Example 3a is

$$P = \begin{array}{c} \\ \mathbf{s}_1 \\ \mathbf{s}_2 \\ \mathbf{s}_3 \\ \mathbf{s}_4 \\ \mathbf{s}_5 \end{array}\!\!\begin{array}{c} \begin{array}{ccccc} \mathbf{s}_1 & \mathbf{s}_2 & \mathbf{s}_3 & \mathbf{s}_4 & \mathbf{s}_5 \end{array} \\ \begin{pmatrix} 0 & 0 & 1 & 0 & 0 \\ {}^1/_2 & 0 & {}^1/_2 & 0 & 0 \\ 0 & {}^1/_2 & 0 & {}^1/_2 & 0 \\ 0 & 0 & {}^1/_2 & 0 & {}^1/_2 \\ 0 & 0 & 1 & 0 & 0 \end{pmatrix} \end{array}.$$

The transition matrix for Example 7 is

$$P = \begin{array}{c} \\ \mathbf{s}_1 \\ \mathbf{s}_2 \\ \mathbf{s}_3 \\ \mathbf{s}_4 \\ \mathbf{s}_5 \end{array}\!\!\begin{array}{c} \begin{array}{ccccc} \mathbf{s}_1 & \mathbf{s}_2 & \mathbf{s}_3 & \mathbf{s}_4 & \mathbf{s}_5 \end{array} \\ \begin{pmatrix} .1 & .2 & .4 & .2 & .1 \\ .1 & .2 & .4 & .2 & .1 \\ .1 & .2 & .4 & .2 & .1 \\ .1 & .2 & .4 & .2 & .1 \\ .1 & .2 & .4 & .2 & .1 \end{pmatrix} \end{array}.$$

The limiting vector for each of these chains is $\alpha = (.1, .2, .4, .2, .1)$. Thus by the Law of Large Numbers we can expect in each case about .1 of the steps to be to s_1, about .2 to s_2, etc.

For a regular Markov chain the fundamental matrix is given by $Z = (I - P + A)^{-1}$. If this is computed for Example 3a we obtain

$$Z = \begin{matrix} & \begin{matrix} s_1 & s_2 & s_3 & s_4 & s_5 \end{matrix} \\ \begin{matrix} s_1 \\ s_2 \\ s_3 \\ s_4 \\ s_5 \end{matrix} & \begin{pmatrix} .88 & -.04 & .32 & -.04 & -.12 \\ .33 & .86 & .12 & -.14 & -.17 \\ -.02 & .16 & .72 & .16 & -.02 \\ -.17 & -.14 & .12 & .86 & .33 \\ -.12 & -.04 & .32 & -.04 & .88 \end{pmatrix} \end{matrix}.$$

For an independent trials process, Z is the identity matrix. Hence for Example 7, $Z = I$.

The first information we obtain from Z relates to the number of times in a state in the first n steps. Let $\bar{\mathbf{y}}^{(n)}{}_j$ be the number of times in state s_j in the first n steps (counting the initial state). Then by Theorem 4.3.4

$$\mathbf{M}_i[\bar{\mathbf{y}}^{(n)}{}_j] - na_j \rightarrow z_{ij} - a_j.$$

For the independent case this limit is replaced by equality. In the dependent case, z_{ij} gives us a comparison of $\mathbf{M}_i[\bar{\mathbf{y}}^{(n)}{}_j]$ for fixed s_j and different starting states s_i. For example, in Example 3a, $z_{11} > z_{21} > z_{31} > z_{51} > z_{41}$. Thus for large n

$$\mathbf{M}_1[\bar{\mathbf{y}}^{(n)}{}_1] > \mathbf{M}_2[\bar{\mathbf{y}}^{(n)}{}_1] > \mathbf{M}_3[\bar{\mathbf{y}}^{(n)}{}_1] > \mathbf{M}_5[\bar{\mathbf{y}}^{(n)}{}_1] > \mathbf{M}_4[\bar{\mathbf{y}}^{(n)}{}_1].$$

The fact that the process may be expected to be more often in s_1 starting from s_5 than from s_4 may be seen also from the fact that to reach s_1 from either of these states it is necessary to go through s_3. From s_5 the first step is to s_3 while from s_4 it is either to s_3 or to s_5.

We can also find from Z the limiting variance for $\bar{\mathbf{y}}^{(n)}{}_j/\sqrt{n}$. This is given by

$$\beta = \{a_j(2z_{jj} - 1 - a_j)\}.$$

In Example 3a this gives

$$\beta = (.066, .104, .016, .104, .066).$$

For the independent trials process $\beta = \{a_j(1 - a_j)\}$. Thus for Example 7,

$$\beta = (.09, .16, .24, .16, .09).$$

The variance in the independent case is in each case larger than the corresponding variance in the dependent case. The variance for s_3 is much larger. This means that we can make more accurate predictions about $\bar{\mathbf{y}}^{(n)}{}_3$ in Example 3a than in Example 7. For example, the

Central Limit Theorem tells us that in 1000 steps the number of occurrences of $\mathbf{s}_3$ in Example 3a will, with probability $\approx .95$, not deviate from 400 by more than $2\sqrt{1000 \cdot .016} = 8$. In Example 7 we could, with the same probability, only say that the number of occurrences of $\mathbf{s}_3$ would deviate from 400 by less than $2\sqrt{1000 \cdot .24} \approx 31$.

Let us next compute the covariance matrix and some correlations. For Example 3a we have

$$C = \begin{pmatrix} .066 & .042 & -.016 & -.058 & -.034 \\ .042 & .104 & .008 & -.096 & -.058 \\ -.016 & .008 & .016 & .008 & -.016 \\ -.058 & -.096 & .008 & .104 & .042 \\ -.034 & -.058 & -.016 & .042 & .066 \end{pmatrix}.$$

For the limiting correlation between $\mathbf{s}_1$ and each of the five states we have (rounded): (1.00, .51, $-.49$, $-.70$, $-.52$).

For Example 7 the covariance matrix is

$$C = \begin{pmatrix} .09 & -.02 & -.04 & -.02 & -.01 \\ -.02 & .16 & -.08 & -.04 & -.02 \\ -.04 & -.08 & .24 & -.08 & -.04 \\ -.02 & -.04 & -.08 & .16 & -.02 \\ -.01 & -.02 & -.04 & -.02 & .09 \end{pmatrix}.$$

The limiting correlations with $\mathbf{s}_1$ are: (1.00, $-.17$, $-.28$, $-.17$, $-.11$).

It is to be expected that often the limiting correlations between two different states will be negative, since—generally—the more often the process enters one state, the less often it will be in the other state. For the independent process all the correlations between pairs of different states are negative, though quite small. But for Example 3a the correlations are fairly large, and the correlation between $\mathbf{s}_1$ and $\mathbf{s}_2$ is positive.

We next consider the function $\mathbf{f}_j$ which gives the number of steps taken to reach $\mathbf{s}_j$ for the first time. The values of $\mathbf{M}_i[\mathbf{f}_j]$ are given by the matrix $M = (I - Z + EZ_{dg})D$. For Example 3a this is

$$M = \begin{matrix} \\ \mathbf{s}_1 \\ \mathbf{s}_2 \\ \mathbf{s}_3 \\ \mathbf{s}_4 \\ \mathbf{s}_5 \end{matrix} \begin{matrix} \begin{matrix} \mathbf{s}_1 & \mathbf{s}_2 & \mathbf{s}_3 & \mathbf{s}_4 & \mathbf{s}_5 \end{matrix} \\ \begin{pmatrix} 10 & 4.5 & 1 & 4.5 & 10 \\ 5.5 & 5 & 1.5 & 5 & 10.5 \\ 9 & 3.5 & 2.5 & 3.5 & 9 \\ 10.5 & 5 & 1.5 & 5 & 5.5 \\ 10 & 4.5 & 1 & 4.5 & 10 \end{pmatrix} \end{matrix}.$$

For an independent trials process the formula for M reduces to $M = ED$. Thus for Example 7

$$M = \begin{array}{c} s_1 \\ s_2 \\ s_3 \\ s_4 \\ s_5 \end{array} \overset{\begin{array}{ccccc} s_1 & s_2 & s_3 & s_4 & s_5 \end{array}}{\begin{pmatrix} 10 & 5 & 2.5 & 5 & 10 \\ 10 & 5 & 2.5 & 5 & 10 \\ 10 & 5 & 2.5 & 5 & 10 \\ 10 & 5 & 2.5 & 5 & 10 \\ 10 & 5 & 2.5 & 5 & 10 \end{pmatrix}}.$$

In the independent case the mean time to reach s_j is independent of the starting state. This is not true for the dependent case. In fact, the mean time required to reach s_1 from s_2 is only about half that required for any other starting state. We observe that the mean time to return to a state, $\mathbf{M}_i[\mathbf{f}_i]$, is the same in the two examples. This is because these means depend only on α.

The variances $\mathbf{Var}_i[\mathbf{f}_j]$ are given by

$$M_2 = M(2Z_{dg}D - I) + 2(ZM - E(ZM)_{dg}) - M_{sq}.$$

For Example 3a this is

$$M_2 = \begin{array}{c} s_1 \\ s_2 \\ s_3 \\ s_4 \\ s_5 \end{array} \overset{\begin{array}{ccccc} s_1 & s_2 & s_3 & s_4 & s_5 \end{array}}{\begin{pmatrix} 66 & 12^3/_4 & 0 & 12^3/_4 & 66 \\ 53^1/_4 & 13 & ^1/_4 & 13 & 66^1/_4 \\ 66 & 12^3/_4 & ^1/_4 & 12^3/_4 & 66 \\ 66^1/_4 & 13 & ^1/_4 & 13 & 53^1/_4 \\ 66 & 12^3/_4 & 0 & 12^3/_4 & 66 \end{pmatrix}}.$$

For the independent trials case the formula for M_2 reduces to $M_2 = E(D^2 - D)$. Thus for Example 7 we have

$$M_2 = \begin{array}{c} s_1 \\ s_2 \\ s_3 \\ s_4 \\ s_5 \end{array} \overset{\begin{array}{ccccc} s_1 & s_2 & s_3 & s_4 & s_5 \end{array}}{\begin{pmatrix} 90 & 20 & ^{15}/_4 & 20 & 90 \\ 90 & 20 & ^{15}/_4 & 20 & 90 \\ 90 & 20 & ^{15}/_4 & 20 & 90 \\ 90 & 20 & ^{15}/_4 & 20 & 90 \\ 90 & 20 & ^{15}/_4 & 20 & 90 \end{pmatrix}}.$$

As in the case of the means, in the independent case, Example 7, the variances do not depend on the starting state. Unlike the case of the means this is almost true for the variances in the dependent case

Example 3a. We note finally that, as in the case of the variances for $\bar{\mathbf{y}}^{(n)}{}_j$, the variances for $\mathbf{f}_j$ are in each case greater for the independent case than for the dependent.

§ 4.8 **The general two-state case.** In this section we give for future reference the basic quantities for Example 11 of Chapter II. We recall that this was the general Markov chain with two states. The transition matrix was written in the form

$$P = \begin{pmatrix} 1-c & c \\ d & 1-d \end{pmatrix}.$$

We assume that $0 < c \leqslant 1$ and $0 < d \leqslant 1$ but c and d are not both 1. This will give us the general regular two-state Markov chain.

The limiting vector α is

$$\alpha = \left(\frac{d}{c+d}, \frac{c}{c+d}\right).$$

The fundamental matrix $Z = (I - P + A)^{-1}$ is

$$Z = \frac{1}{c+d}\begin{pmatrix} d + \dfrac{c}{c+d} & c - \dfrac{c}{c+d} \\ d - \dfrac{d}{c+d} & c + \dfrac{d}{c+d} \end{pmatrix}.$$

The mean first passage matrix M is

$$M = \begin{pmatrix} \dfrac{c+d}{d} & \dfrac{1}{c} \\ \dfrac{1}{d} & \dfrac{c+d}{c} \end{pmatrix},$$

and the variance matrix for the first passage time is

$$M_2 = \begin{pmatrix} \dfrac{c(2-c-d)}{d^2} & \dfrac{1-c}{c^2} \\ \dfrac{1-d}{d^2} & \dfrac{d(2-c-d)}{c^2} \end{pmatrix}.$$

The limiting variance for the number of times in state s_j is given by

$$\beta = \left(\frac{cd(2-c-d)}{(c+d)^3}, \frac{cd(2-c-d)}{(c+d)^3}\right).$$

Compare this variance for the dependent case with the independent case having the same limiting vector. This independent process would have transition matrix

$$P = \begin{pmatrix} \frac{d}{c+d} & \frac{c}{c+d} \\ \frac{d}{c+d} & \frac{c}{c+d} \end{pmatrix},$$

and the limiting variance for the number of times in s_j would be

$$\beta = \left(\frac{cd}{(c+d)^2}, \quad \frac{cd}{(c+d)^2} \right).$$

Thus the limiting variance for the number of times in s_1 will be greater in the dependent case if and only if

$$2-c-d > c+d.$$

That is, if the sum of the diagonal elements is greater than the sum of the off-diagonal elements. Or in other words, if the probabilities for remaining in a state have a sum greater than the probabilities for a change of state.

The covariance matrix is

$$C = \frac{cd(2-c-d)}{(c+d)^3} \begin{pmatrix} 1 & -1 \\ -1 & 1 \end{pmatrix}.$$

Thus $c_{ij} > 0$ if $i=j$, but $c_{ij} < 0$ if $i \neq j$. The limiting correlations are $+1$ and -1 in the two cases, respectively.

Exercises for Chapter IV

For § 4.1

1. Find the limiting matrix A for Example 13. (See Exercise 23, Chapter II.)

2. Find the limiting matrix A for Example 14. (See Exercise 24, Chapter II.)

3. Show that the four-state chain in Example 12 is regular. Find the fixed vector α. What is the relation of this vector to the fixed vector for the two-state chain which determined the four-state chain?

4. Show that if α is the fixed probability vector for a chain with transition matrix P, then it is also a fixed vector for the chain with transition matrix P^n.

5. Prove that if a transition matrix has column sums 1, then the fixed vector has equal components.

6. Given a probability vector α with positive components, determine a regular transition matrix which will have this as its fixed vector.

For § 4.2

7. For Example 14 find the mean and variance for the number of times in state s_1 in the first n steps.

8. Consider the Markov chain with transition matrix

$$P = \begin{matrix} & s_1 & s_2 \\ s_1 & 0 & 1 \\ s_2 & 1/2 & 1/2 \end{matrix}.$$

Start the process in s_2, and compute the mean of $\mathbf{v}^{(n)}_1$ for $n=1, 2, 3, 4, 5, 6$. Compare these results with a_1.

For § 4.3

9. Find the fundamental matrix for Example 11 when $c={}^1/_2$ and $d={}^1/_4$.

10. Find the limit of the difference between the mean number of nice days in the Land of Oz in the first n days, starting with a rainy day and starting with a nice day.

11. Find the fundamental matrix for the chain in Exercise 8 above. Interpret $z_{11}-z_{21}$.

12. Find the fundamental matrix for Example 14. (Use the result of Exercise 2 above.)

13. Find the fundamental matrix for Example 13. (Use the result of Exercise 1 above.)

For § 4.4

14. Find the mean first passage matrix for Example 14. (Use the result of Exercise 12 above.)

15. Find the mean first passage matrix for Example 13. (Use the result of Exercise 13 above).

16. Verify Theorems 4.4.9 and 4.4.10 for Example 13.

17. Prove that for an independent trials process, M has all rows the same.

18. Given that the mean first passage matrix of a chain has the form

$$M = \begin{pmatrix} x & 2 & 6 \\ 4 & x & 4 \\ 6 & 2 & x \end{pmatrix}$$

determine the transition matrix.

19. Give two different transition matrices which have the same fundamental matrix, and hence show that the fundamental matrix does not determine the transition matrix.

20. Prove that P has constant column sums if and only if $\bar{M}$ has constant row-sums.

For § 4.5

21. Find M_2 for Example 14.

22. Using the result of Exercise 15 above, find M_2 for Example 13.

23. Find M_2 for Example 11 with $c={}^1/_2$ and $d={}^1/_4$.

24. A die is rolled a number of times. Find the mean and variance for the number of rolls between occurrences of two 6's.

25. Find the mean and variance of the first passage times in Exercise 8 above.

For § 4.6

26. Find the limiting covariance matrix for Example 11 with $c={}^1/_2$ and $d={}^1/_4$.

27. Find the limiting covariance matrix for Example 13. Interpret the diagonal entries.

28. On a nice day a man in the Land of Oz takes his umbrella with probability ${}^1/_2$, on a rainy day with probability 1 and on a snowy day with probability ${}^3/_4$. Find the limiting variance for the number of days that he will take his umbrella.

29. For an absorbing chain let $\mathbf{n}_j$ be the number of times in state s_j before absorption. Using the method of proof for Theorem 4.6.1, show that

$$M_k[\mathbf{n}_i \cdot \mathbf{n}_j] = n_{kj}n_{ji} + n_{ki}n_{ij} - d_{ij}n_{ki}$$

where $N=\{n_{ij}\}$ is the fundamental matrix. Find $\mathbf{Cov}_k[\mathbf{n}_i,\mathbf{n}_j]$ and $\mathbf{Var}_k[\mathbf{n}_i]$.

For § 4.8

30. Find the limiting variance of the number of times in a state when $c=d$. How does this vary with c? Interpret your formula as $c \to 0$.

31. Find the limiting vector and the mean first passage matrix for the case where $c=2d$. How do these vary with c? Interpret your results as $c \to 0$.

For the entire chapter

32. Consider the following transition matrix for a Markov chain.

$$P = \begin{pmatrix} {}^1/_2 & {}^1/_3 & {}^1/_6 \\ {}^3/_4 & 0 & {}^1/_4 \\ 0 & 1 & 0 \end{pmatrix} \begin{matrix} s_1 \\ s_2 \\ s_3 \end{matrix}$$

(a) Is the chain regular?
(b) Find α, A, and Z.
(c) Find M and M_2.
(d) Find the covariance matrix.
(e) Use absorbing-chain methods to find the mean time to go from s_3 to s_1. Check your answer against part (c).

33. Let P_1 and P_2 be two different transition matrices for a three-state

Markov chain. By a random device we select one of these matrices and carry out the resulting chain. (Say P_1 is selected with probability p.)

(a) Is this process a Markov chain?
(b) Show that the probability of being in a given state tends to a limit, and show how these probabilities may be obtained from the fixed vectors of the two matrices.

34. Suppose that in Exercise 33 we use the random device before each step, to decide which matrix to apply on that step.

(a) Is this process a Markov chain?
(b) Show that the limiting probabilities for being in the various states are normally not the same as those obtained in Exercise 33(b).

CHAPTER V

ERGODIC MARKOV CHAINS

§ **5.1 Fundamental matrix.** We will now generalize the results obtained in the last chapter. There they were proved for regular chains, and now we will extend them to an arbitrary chain consisting of a single ergodic set, i.e. to an ergodic chain. We know that such a chain must be either regular or cyclic. A cyclic chain consists of d cyclic classes, and a regular chain may be thought of as the special case where $d=1$. The results to be obtained will be generalizations of the previous results in the sense that if we set $d=1$ in them, we obtain a result from the previous chapter. As a matter of fact, in most of the results d will not appear explicitly, so that the result of the previous chapter will be shown to hold for all ergodic chains.

An ergodic chain is characterized by the fact that it consists of a single ergodic class, that is, it is possible to go from every state to every other state. However, if $d>1$, then such transition is possible only for special n-values. Thus no power of P is positive, and different powers will have zeros in different positions, these zeros changing cyclically for the powers. Hence P^n cannot converge. This is the most important difference between cyclic and regular chains.

But while the powers fail to converge, we have the following weaker result.

5.1.1 THEOREM. *For any ergodic chain the sequence of powers P^n is Euler-summable to a limiting matrix A, and this limiting matrix is of the form $A=\xi\alpha$, with α a positive probability vector.*

PROOF. Consider the matrix $(kI+(1-k)P)$, for some k, $0<k<1$. This matrix is again a transition matrix. Since it has positive entries in all places where P is positive, the new matrix also represents an ergodic chain. And since the diagonal entries are positive, it is possible to return to a state in one step, and hence $d=1$. Thus the new chain is regular.

From § 4.1.4 we know that $(kI+(1-k)P)^n$ tends to a matrix $A=\xi\alpha$, with a probability vector $\alpha>0$. Thus

$$A = \lim_{n\to\infty} (kI+(1-k)P)^n$$

$$A = \lim_{n\to\infty} \sum_{i=0}^{n} \binom{n}{i} k^{n-i}(1-k)^i P^i. \qquad (1)$$

But this states precisely that the sequence P^n is Euler-summable to A (see § 1.10). Indeed, it is Euler-summable for every value of k.

5.1.2 THEOREM. *If P is an ergodic transition matrix, and A and α are as in Theorem* 5.1.1, *then*

(a) *For any probability vector π, the sequence πP^n is Euler-summable to α.*
(b) *The vector α is the unique fixed probability vector of P.*
(c) $PA=AP=A$,

PROOF. If we multiply (1) by π we obtain that the Euler sum of the sequence πP^n is $\pi A=\pi\xi\alpha=\alpha$, which proves (a).

Since α was obtained from the limiting matrix of $(kI+(1-k)P)$, it is the unique fixed probability vector of this regular transition matrix. But this matrix must have the same fixed vectors as P, since

$$\pi(kI+(1-k)P) = \pi$$

implies that

$$\pi(1-k)P = \pi(1-k)$$

and since $k\neq 1$,

$$\pi P = \pi.$$

This proves (b). Part (c) follows from the fact that $P\xi=\xi$ for any transition matrix, and that $\alpha P=\alpha$.

We thus see that α and A have nearly the same properties in the ergodic case as they did for regular chains; only, (a) had to be weakened to summability in place of convergence. We will now show that ergodic chains have a fundamental matrix which behaves just like the fundamental matrix of regular chains.

5.1.3 THEOREM. *If P is an ergodic transition matrix, then the inverse matrix $Z=(I-(P-A))^{-1}$ exists, and*

(a) $PZ = ZP$
(b) $Z\xi = \xi$
(c) $\alpha Z = \alpha$
(d) $(I-P)Z = I-A$.

PROOF. Since the sequence P^n is Euler-summable to A by § 5.1.1, and since $(P-A)^n = P^n - A$ by § 5.1.2(c), the sequence $(P-A)^n$ is Euler-summable to 0. Hence the inverse Z exists (see § 1.11). Furthermore, the series

$$I + \sum_{i=1}^{\infty} (P^i - A) \tag{2}$$

is Euler-summable to Z. Then (b) and (c) follow from the fact that $I\xi = \xi$, $\alpha I = \alpha$, and multiplying $P^i - A$ by either ξ on the right or by α on the left yields 0. Result (d) is obtained by multiplying (2) by $I-P$.

While for the theorems so far, Euler-summability of P^n sufficed, we will need the following stronger results.

5.1.4 THEOREM. *If P is an ergodic transition matrix,*

(a) *The sequence P^n is Cesaro-summable to A.*

(b) *The series $I + \sum_{i=1}^{\infty} (P^i - A)$ is Cesaro-summable to Z.*

PROOF. If $n = kd$, then in n steps after starting at s_i we must be in a state in the cyclic class of s_i. And if k is sufficiently large, we may be in any state in the class. Hence P^d may be thought of as the transition matrix of a Markov chain with d separate ergodic sets, each of which is non-cyclic. Therefore, P^{kd} tends to a limiting matrix A_0, whose ij-entry is 0 if s_i and s_j are not in the same cyclic class, and otherwise the ij entry is gotten by taking the components of α belonging to the cyclic class and renormalizing them.

If $0 \leqslant l < d$, then P^{kd+l} tends to $P^l A_0$ as k tends to infinity. Hence the sequence P^n has these d convergent subsequences, and hence (see § 1.10) P^n is Cesaro-summable to the average of the limits. But two different summation methods cannot give different answers, hence A must be this average; that is,

$$A = (1/d) \sum_{l=0}^{d-1} P^l A_0, \tag{3}$$

and P is Cesaro-summable to A. It is then an immediate consequence that since $P^i - A$ is Cesaro-summable to 0, (b) must hold.

Let us restate the summability result (b) as a limit.

5.1.5 COROLLARY. $I + \lim_{n \to \infty} \sum_{i=1}^{n} \frac{n-i}{n} (P^i - A) = Z.$

Since we have now succeeded in generalizing many basic properties of Z to ergodic chains, and since d did not appear explicitly, we may now assert that many of the results of Chapter IV hold for all ergodic chains. In particular this applies to all results concerning the mean first passage time matrix M and the variance of first passage time matrix M_2, that is to all results in §§ 4.4 and 4.5. We also have all the results of § 4.6 concerning limiting variances and covariances, since in the proof of § 4.6.1 we needed only the summability (5.1.5) of the infinite series for Z, not its convergence. And thus all the basic formulas of §§ 4.4, 4.5, and 4.6 may be applied to any ergodic chain.

It is worth making a special comment on § 4.4.12. We now know that $\bar{M}$ determines the transition matrix of any ergodic chain by means of the formula $P = I + (D - E)\bar{M}^{-1}$. Thus, in particular, $\bar{M}$ determines whether (or not) the chain is cyclic. This is quite surprising, and it would be highly desirable to find necessary and sufficient conditions (i) that $\bar{M}$ be the mean first passage matrix of an ergodic chain, and (ii) that it represent a regular rather than a cyclic chain. One would like these conditions to be simpler than computing P from $\bar{M}$ and then checking P.

What results on regular chains have we not generalized so far? The most important such results are: The geometric estimate § 4.1.5, the Law of Large Numbers § 4.2.1, and the results in §§ 4.3.4–4.3.6 on $\bar{\mathrm{y}}^{(n)}{}_j$. To be able to discuss these we shall have to find some sort of an upper bound on $p^{(n)}{}_{ij} - a_j$.

It is clear that the geometric bound of § 4.1.5 cannot apply to this difference in the cyclic chain, since $p^{(n)}{}_{ij}$ will frequently be 0, and hence the difference—in absolute value—is a_j infinitely often. However it can be shown, using the ideas of the proof of § 5.1.4, that if we add up d consecutive terms, that is, form

$$\sum_{l=0}^{d-1} (p^{(n+l)}{}_{ij} - a_j),$$

then this sum is bounded geometrically. This suffices to prove the Law of Large Numbers, if in § 4.2.1 we take sums d terms at a time. This method also allows us to prove analogues of §§ 4.3.4–4.3.6, but we will not take these up.

§ 5.2 Examples of cyclic chains. The simplest possible cyclic chain is obtained from the two-state Example 11 by choosing $c = d = 1$. We will call this Example 11a. The transition matrix is

$$P = \begin{pmatrix} 0 & 1 \\ 1 & 0 \end{pmatrix}.$$

From the proof of § 5.1.1 we know that A may be obtained as the limiting matrix of $(^1/_2)I + (^1/_2)P = (^1/_2)E$. But this is its own limiting matrix. Hence

$$A = (^1/_2)E, \qquad \alpha = (^1/_2, ^1/_2), \qquad d = 2,$$

$$Z = \begin{pmatrix} ^3/_2 & -^1/_2 \\ -^1/_2 & ^3/_2 \end{pmatrix}^{-1} = \begin{pmatrix} ^3/_4 & ^1/_4 \\ ^1/_4 & ^3/_4 \end{pmatrix}.$$

$$M = \begin{pmatrix} 2 & 1 \\ 1 & 2 \end{pmatrix}, \qquad M_2 = \begin{pmatrix} 0 & 0 \\ 0 & 0 \end{pmatrix}.$$

It is very easy to find M directly, and to see that M_2 must have all components 0. Similarly, the limiting variances are 0.

As a less trivial example we take up the random walk Example 2 for $p = ^1/_2$ (denoted by Example 2a). Its transition matrix is

$$P = \begin{matrix} s_1 \\ s_2 \\ s_3 \\ s_4 \\ s_5 \end{matrix} \begin{pmatrix} 0 & 1 & 0 & 0 & 0 \\ ^1/_2 & 0 & ^1/_2 & 0 & 0 \\ 0 & ^1/_2 & 0 & ^1/_2 & 0 \\ 0 & 0 & ^1/_2 & 0 & ^1/_2 \\ 0 & 0 & 0 & 1 & 0 \end{pmatrix}.$$

Starting from an even-numbered state, the process can be in even-numbered states only in an even number of steps, and in an odd-numbered state in an odd number of steps; hence the even and odd states form two cyclic classes. Computing the other quantities we find:

$$\alpha = (^1/_8, ^1/_4, ^1/_4, ^1/_4, ^1/_8)$$

$$Z = {^1/_{16}} \begin{pmatrix} 23 & 18 & -2 & -14 & -9 \\ 9 & 22 & 2 & -10 & -7 \\ -1 & 2 & 14 & 2 & -1 \\ -7 & -10 & 2 & 22 & 9 \\ -9 & -14 & -2 & 18 & 23 \end{pmatrix}.$$

$$M = \begin{matrix} \\ s_1 \\ s_2 \\ s_3 \\ s_4 \\ s_5 \end{matrix} \begin{matrix} \begin{matrix} s_1 & s_2 & s_3 & s_4 & s_5 \end{matrix} \\ \begin{pmatrix} 8 & 1 & 4 & 9 & 16 \\ 7 & 4 & 3 & 8 & 15 \\ 12 & 5 & 4 & 5 & 12 \\ 15 & 8 & 3 & 4 & 7 \\ 16 & 9 & 4 & 1 & 8 \end{pmatrix} \end{matrix}.$$

$$M_2 = \begin{matrix} s_1 \\ s_2 \\ s_3 \\ s_4 \\ s_5 \end{matrix} \begin{pmatrix} 112 & 0 & 8 & 48 & 160 \\ 112 & 24 & 8 & 48 & 160 \\ 152 & 40 & 8 & 40 & 152 \\ 160 & 48 & 8 & 24 & 112 \\ 160 & 48 & 8 & 0 & 112 \end{pmatrix}.$$

It is interesting to examine some of the entries of M and of M_2. From either end state we can go to any state only by passing through the neighboring state. Hence the first row of M is, with one exception, 1 greater than the second row, and similarly for the fifth and fourth rows. The one exception is stepping into the neighboring state itself. The third row is the average of the second and fourth, plus 1, except for stepping into a neighboring state.

In M_2 it is worth noting the equal entries. Some of the equalities are due to the symmetry of the process. But this does not account, for example, for the third column being constant. The second and fourth entries are the same in this column by symmetry. The other three are also 8, because from one of the states in the first cyclic set we must enter the second cyclic set, and then the variance is 8. The two 0 entries are due to the fact that from an end state we always go to its neighbor in one step.

It is also interesting to think of the middle column in M and M_2 as arising from making s_3 absorbing, and asking for the mean and variance of the number of steps needed for absorption. The resulting process behaves in all essential features like § 3.4.1 (with $p = 1/2$), and hence the numbers 3, 4, and 8 are the same as the entries of τ and τ_2 there obtained.

We shall conclude by computing the covariance matrix.

$$C = 1/32 \begin{pmatrix} 7 & 8 & -2 & -8 & -5 \\ 8 & 12 & 0 & -12 & -8 \\ -2 & 0 & 4 & 0 & -2 \\ -8 & -12 & 0 & 12 & 8 \\ -5 & -8 & -2 & 8 & 7 \end{pmatrix}.$$

From this we obtain the limiting variances

$$\beta = (7/32, 3/8, 1/8, 3/8, 7/32).$$

The fact that $c_{23} = c_{43} = 0$ means that the limiting correlations between s_2 and s_3, and s_4 and s_3, are 0. On the other hand the correlation between s_1 and s_2 is $8/\sqrt{84} \approx .87$. The reason for this is fairly obvious from the transition matrix.

§ **5.3 Reverse Markov chains.** We saw in § 2.1 that a Markov process observed in reverse order would be a Markov process with transition probabilities given by

$$p_{ij}(n) = \frac{\mathbf{Pr}_\pi[\mathbf{f}_{n-1}=\mathbf{s}_j]\mathbf{Pr}_\pi[\mathbf{f}_n=\mathbf{s}_i|\mathbf{f}_{n-1}=\mathbf{s}_j]}{\mathbf{Pr}_\pi[\mathbf{f}_n=\mathbf{s}_i]}$$

where $\mathbf{f}_n$ is the n-th outcome function. It was observed also that, if the forward process is a Markov chain, the reverse process will be a Markov chain only if $\mathbf{Pr}_\pi[\mathbf{f}_n=\mathbf{s}_j]$ does not depend on n. This will be the case if the process is started in equilibrium. In this case $\mathbf{Pr}_\alpha[\mathbf{f}_n=\mathbf{s}_i]=a_i$ for all n, and $p_{ij}(n)$ becomes

$$\hat{p}_{ij} = p_{ij}(n) = \frac{a_j p_{ji}}{a_i}.$$

5.3.1 DEFINITION. *Let P be the transition matrix for an ergodic Markov chain. Let α be the fixed probability vector for P. Then the* reverse Markov chain *for P is a Markov chain with transition matrix given by*

$$\hat{P} = \{\hat{p}_{ij}\} = \left\{\frac{a_j p_{ji}}{a_i}\right\} = DP^TD^{-1}.$$

To justify the above definition we must show that P is a transition matrix. By Theorem 5.1.2 the a_i's are all positive, so that $\hat{p}_{ij}$ is defined and non-negative.

$\hat{P}\xi = DP^TD^{-1}\xi = DP^T\alpha^T = D(\alpha P)^T = D\alpha^T = \xi$. Hence $\hat{P}$ is a transition matrix.

5.3.2 DEFINITION. *A Markov chain is* reversible *if $P=\hat{P}$.*

5.3.3 THEOREM. *A Markov chain is reversible if and only if $D^{-1}P$ is a symmetric matrix.*

PROOF. $\hat{P}=DP^TD^{-1}$. Hence $P=\hat{P}$ if and only if

$$D^{-1}P = P^TD^{-1} = (D^{-1}P)^T.$$

That is, if and only if $D^{-1}P$ is a symmetric matrix.

A reversible Markov chain in equilibrium will appear the same looked at backwards as forwards. An alternative way to describe reversibility is the following. A process is reversible if, in equilibrium, for any $\mathbf{s}_i$ and $\mathbf{s}_j$ the probability of $\mathbf{s}_i$ followed by $\mathbf{s}_j$ is the same as the probability of $\mathbf{s}_j$ followed by $\mathbf{s}_i$. That is, if for every n, $\mathbf{s}_i$, $\mathbf{s}_j$

$$\mathbf{Pr}_\alpha[\mathbf{f}_n=\mathbf{s}_i \wedge \mathbf{f}_{n+1}=\mathbf{s}_j] = \mathbf{Pr}_\alpha[\mathbf{f}_n=\mathbf{s}_j \wedge \mathbf{f}_{n+1}=\mathbf{s}_i].$$

This last equation will be true if $a_ip_{ij}=a_jp_{ji}$ or if $p_{ij}=a_jp_{ji}/a_i$. That is, if $p_{ij}=\hat{p}_{ij}$ for every i, j.

It is obvious that any periodic chain with period greater than 2 cannot be reversible. In fact for such a chain, any state which can be reached on the next step could not have been the result of the last step. Thus only chains of period 1 and 2 can be reversible. It is clear that if such a chain is reversible it will have the same period.

As an example of a reversible chain of period 1, we can consider the Land of Oz example. In this case we find:

$$D^{-1}P = \begin{pmatrix} 2/5 & 0 & 0 \\ 0 & 1/5 & 0 \\ 0 & 0 & 2/5 \end{pmatrix} \begin{pmatrix} 1/2 & 1/4 & 1/4 \\ 1/2 & 0 & 1/2 \\ 1/4 & 1/4 & 1/2 \end{pmatrix}$$

$$= \begin{array}{c} \\ \mathbf{R} \\ \mathbf{N} \\ \mathbf{S} \end{array} \begin{array}{c} \begin{array}{ccc} \mathbf{R} & \mathbf{N} & \mathbf{S} \end{array} \\ \begin{pmatrix} 1/5 & 1/10 & 1/10 \\ 1/10 & 0 & 1/10 \\ 1/10 & 1/10 & 1/5 \end{pmatrix} \end{array},$$

which is a symmetric matrix. Hence by Theorem 5.3.3 the chain is reversible. An example of a reversible chain with period 2 is given by Example 2a. In this case the matrix $D^{-1}P$ is

$$D^{-1}P = \begin{pmatrix} 0 & 1/8 & 0 & 0 & 0 \\ 1/8 & 0 & 1/8 & 0 & 0 \\ 0 & 1/8 & 0 & 1/8 & 0 \\ 0 & 0 & 1/8 & 0 & 1/8 \\ 0 & 0 & 0 & 1/8 & 0 \end{pmatrix}.$$

This is again a symmetric matrix.

Given an ergodic chain, we now ask for the relation between this chain and the associated reverse chain. We shall find the relation between the fixed vectors and the fundamental matrices and hence, from these, any quantities which depend on them. We shall denote A, Z, M, etc. for the reverse chain by $\hat{A}$, $\hat{Z}$, $\hat{M}$, etc.

5.3.4 THEOREM. *The fixed probability vector for P and $\hat{P}$ is the same.*

PROOF. Let
$$\alpha P = \alpha.$$
Then
$$\begin{aligned} \alpha\hat{P} &= \alpha D P^T D^{-1} \\ &= \eta P^T D^{-1} \\ &= (P\xi)^T D^{-1} \\ &= \eta D^{-1} \\ &= \alpha. \end{aligned}$$

5.3.5 THEOREM. $\hat{Z} = DZ^T D^{-1}$.

PROOF. $$\hat{Z} = (I - \hat{P} + \hat{A})^{-1}.$$

From the form of A it is clear that $A = DA^T D^{-1}$. From Theorem 5.3.4, $\hat{A} = A$. Thus

$$\begin{aligned}\hat{Z} &= (I - DP^T D^{-1} + DA^T D^{-1})^{-1} \\ &= D(I - P^T + A^T)^{-1} D^{-1} \\ &= DZ^T D^{-1}.\end{aligned}$$

5.3.6 THEOREM. *Any quantity whose value depends only on Z_{dg} and A is the same for the reverse process as for the forward process.*

PROOF. By Theorem 5.3.5, $\hat{Z}_{dg} = Z_{dg}$, and by § 5.3.4, $\hat{A} = A$.

An example of the application of the above theorem is the mean and variance of the first passage time to state s_i, if we start in s_i or if we have as initial vector α. Similarly, the limiting variance for the number of times in a state depends only on Z_{dg} and A. Hence these quantities are the same for the forward and reverse processes. Additional examples are provided by the following theorem.

5.3.7 THEOREM. $\hat{C} = C$.

PROOF.
$$\begin{aligned}\hat{c}_{ij} &= a_i \hat{z}_{ij} + a_j \hat{z}_{ji} - a_i d_{ij} - a_i a_j \\ &= a_i(a_j z_{ji}/a_i) + a_j(a_i z_{ij}/a_j) - a_i d_{ij} - a_i a_j \\ &= a_j z_{ji} + a_i z_{ij} - a_i d_{ij} - a_i a_j \\ &= c_{ij}.\end{aligned}$$

Thus all results that depend only on the covariance matrix are the same for the reverse process.

5.3.8 THEOREM. $\hat{M} - M = (ZD) - (ZD)^T$.

PROOF.
$$\begin{aligned}\hat{M} - M &= (I - \hat{Z} + EZ_{dg})D - (I - Z + EZ_{dg})D \\ &= (Z - \hat{Z})D.\end{aligned}$$

The theorem then follows from § 5.3.5.

5.3.9 THEOREM.
$$\begin{aligned}\hat{W} - W = (ZD - (ZD)^T)(2Z_{dg}D - 3I) \\ + 2(Z^2D - (Z^2D)^T).\end{aligned}$$

PROOF.

$$\hat{W} - W = (\hat{M} - M)(2Z_{dg}D - I) + 2(\hat{Z}\hat{M} - ZM) - 2E(\hat{Z}\hat{M} - ZM)_{dg}. \quad (1)$$

$$\hat{M} - M = ZD - (ZD)^T. \quad (2)$$

Since $\hat{Z}\hat{M} = (\hat{Z} - \hat{Z}^2 + EZ_{dg})D$, and $ZM = (Z - Z^2 + EZ_{dg})D$, we have

$$\begin{aligned} \hat{Z}\hat{M} - ZM &= (\hat{Z} - Z)D + (Z^2 - \hat{Z}^2)D \\ &= (ZD)^T - (ZD) + (Z^2D) - (Z^2D)^T. \end{aligned} \tag{3}$$

Since this is the difference of a matrix and its transpose, it has 0 diagonal entries, hence

$$(\hat{Z}\hat{M} - ZM)_{dg} = 0. \tag{4}$$

We obtain our theorem by combining (1), (2), (3), and (4).

We shall now illustrate the application of the above theorems for a process which is not reversible. Such a process is the random walk Example 3a. Here

$$P = \begin{matrix} & \begin{matrix} s_1 & s_2 & s_3 & s_4 & s_5 \end{matrix} \\ \begin{matrix} s_1 \\ s_2 \\ s_3 \\ s_4 \\ s_5 \end{matrix} & \begin{pmatrix} 0 & 0 & 1 & 0 & 0 \\ 1/2 & 0 & 1/2 & 0 & 0 \\ 0 & 1/2 & 0 & 1/2 & 0 \\ 0 & 0 & 1/2 & 0 & 1/2 \\ 0 & 0 & 1 & 0 & 0 \end{pmatrix} \end{matrix}$$

and $\alpha = (.1, .2, .4, .2, .1)$. From this we find

$$\hat{P} = \begin{matrix} & \begin{matrix} s_1 & s_2 & s_3 & s_4 & s_5 \end{matrix} \\ \begin{matrix} s_1 \\ s_2 \\ s_3 \\ s_4 \\ s_5 \end{matrix} & \begin{pmatrix} 0 & 1 & 0 & 0 & 0 \\ 0 & 0 & 1 & 0 & 0 \\ 1/4 & 1/4 & 0 & 1/4 & 1/4 \\ 0 & 0 & 1 & 0 & 0 \\ 0 & 0 & 0 & 1 & 0 \end{pmatrix} \end{matrix}.$$

Most of the entries in this matrix are obvious. For example, if the process is ever in state s_1 it must have come from s_2, hence $\hat{p}_{12} = 1$. The fixed vector for $\hat{p}$ is again $\alpha = (.1, .2, .4, .2, .1)$.

In Chapter IV we found Z for this example to be

$$Z = \begin{matrix} & \begin{matrix} s_1 & s_2 & s_3 & s_4 & s_5 \end{matrix} \\ \begin{matrix} s_1 \\ s_2 \\ s_3 \\ s_4 \\ s_5 \end{matrix} & \begin{pmatrix} .88 & -.04 & .32 & -.04 & -.12 \\ .33 & .86 & .12 & -.14 & -.17 \\ -.02 & .16 & .72 & .16 & -.02 \\ -.17 & -.14 & .12 & .86 & .33 \\ -.12 & -.04 & .32 & -.04 & .88 \end{pmatrix} \end{matrix}.$$

From this we find $\hat{Z} = DZ^T D^{-1}$ to be

$$\hat{Z} = \begin{array}{c} \\ s_1 \\ s_2 \\ s_3 \\ s_4 \\ s_5 \end{array} \begin{array}{c} \begin{array}{ccccc} s_1 & s_2 & s_3 & s_4 & s_5 \end{array} \\ \begin{pmatrix} .88 & .66 & -.08 & -.34 & -.12 \\ -.02 & .86 & .32 & -.14 & -.02 \\ .08 & .06 & .72 & .06 & .08 \\ -.02 & -.14 & .32 & .86 & -.02 \\ -.12 & -.34 & -.08 & .66 & .88 \end{pmatrix} \end{array}.$$

We found M to be:

$$M = \begin{array}{c} \\ s_1 \\ s_2 \\ s_3 \\ s_4 \\ s_5 \end{array} \begin{array}{c} \begin{array}{ccccc} s_1 & s_2 & s_3 & s_4 & s_5 \end{array} \\ \begin{pmatrix} 10 & 4.5 & 1 & 4.5 & 10 \\ 5.5 & 5 & 1.5 & 5 & 10.5 \\ 9 & 3.5 & 2.5 & 3.5 & 9 \\ 10.5 & 5 & 1.5 & 5 & 5.5 \\ 10 & 4.5 & 1 & 4.5 & 10 \end{pmatrix} \end{array}.$$

From this we obtain $\hat{M} = M + (ZD - (ZD)^T)$:

$$\hat{M} = \begin{array}{c} \\ s_1 \\ s_2 \\ s_3 \\ s_4 \\ s_5 \end{array} \begin{array}{c} \begin{array}{ccccc} s_1 & s_2 & s_3 & s_4 & s_5 \end{array} \\ \begin{pmatrix} 10 & 1 & 2 & 6 & 10 \\ 9 & 5 & 1 & 5 & 9 \\ 8 & 4 & 2.5 & 4 & 8 \\ 9 & 5 & 1 & 5 & 9 \\ 10 & 6 & 2 & 1 & 10 \end{pmatrix} \end{array}.$$

We found W to be

$$W = \begin{array}{c} \\ s_1 \\ s_2 \\ s_3 \\ s_4 \\ s_5 \end{array} \begin{array}{c} \begin{array}{ccccc} s_1 & s_2 & s_3 & s_4 & s_5 \end{array} \\ \begin{pmatrix} 166 & 33 & 1 & 33 & 166 \\ 83.5 & 38 & 2.5 & 38 & 176.5 \\ 147 & 25 & 6.5 & 25 & 147 \\ 176.5 & 28 & 2.5 & 28 & 83.5 \\ 166 & 33 & 1 & 33 & 166 \end{pmatrix} \end{array}.$$

From this we obtain $\hat{W} = W + (ZD - (ZD)^T)(2Z_{dg}D - 3I) + 2(Z^2D - (Z^2D)^T)$:

$$\hat{W} = \begin{pmatrix} 166 & 1 & 4 & 49 & 166 \\ 147 & 38 & 1 & 38 & 147 \\ 130 & 29 & 6.5 & 29 & 130 \\ 147 & 38 & 1 & 38 & 147 \\ 166 & 49 & 4 & 1 & 166 \end{pmatrix}.$$

Hence

$$\hat{M}_2 = \begin{pmatrix} 66 & 0 & 0 & 13 & 66 \\ 66 & 13 & 0 & 13 & 66 \\ 66 & 13 & 1/4 & 13 & 66 \\ 66 & 13 & 0 & 13 & 66 \\ 66 & 13 & 0 & 0 & 66 \end{pmatrix}.$$

The zeros and the equal variances are easily deduced from $\hat{P}$.

Exercises for Chapter V

For § 5.1

1. Compute the limiting matrix A and the fundamental matrix for the ergodic chain with transition matrix

$$P = \begin{pmatrix} 0 & 1 & 0 \\ q & 0 & p \\ 0 & 1 & 0 \end{pmatrix}.$$

2. Compute M and M_2 for the chain in Exercise 1 above.

3. Find the covariance matrix C for the chain in Exercise 1 above.

4. In Example 2 let $p = {}^2/_3$. Find the fixed probability vector and the fundamental matrix.

5. For the example of Exercise 4 above find the mean first passage matrix. Check your results by obtaining P from $\hat{M}$.

6. Given that for an ergodic chain

$$M = \begin{pmatrix} 2 & 3 & 3 \\ 1 & 4 & 4 \\ 1 & 4 & 4 \end{pmatrix},$$

show that the chain is cyclic.

7. Prove that the matrix

$$M = \begin{pmatrix} 3 & 4 & {}^{10}/_3 \\ {}^8/_3 & 3 & {}^8/_3 \\ {}^{10}/_3 & 4 & 3 \end{pmatrix}$$

is *not* the first passage matrix of an ergodic chain.

8. Let P be the transition matrix for an ergodic chain. Let $\bar{P}$ be the matrix P with diagonal entries replaced by 0's and the rows renormalized to have sum 1. Show that the resulting chain is again ergodic; and if $\alpha = \{a_j\}$ is the

fixed vector for the original chain, then $\bar{\alpha}=\{a_j(1-p_{jj})\}$ is proportional to the fixed vector for the new chain. What is the interpretation of the components of the new fixed vector in terms of the original chain?

9. Carry out the procedure indicated in the previous exercise for the Land of Oz example.

For § 5.3

10. Find the reverse transition matrix for the chain in Exercise 1 above. Compute the fundamental matrix for this reverse chain from the fundamental matrix for the original chain.

11. For which values of p is the chain in Exercise 1 reversible?

12. Find the reverse transition matrix for Example 2 with $p={}^2/_3$. Compute the fundamental matrix for the reverse chain and compare your result with the result of Exercise 4 above.

13. Compute $\hat{M}$ for the example of the last exercise directly from the fundamental matrix there found. Compute $\hat{M}$ from M (see Exercise 5) using Theorem 5.3.8, and compare your answers.

14. Prove that every independent process is reversible.

15. Prove that every two-state ergodic chain is reversible.

16. Prove that if an ergodic chain has a symmetric transition matrix (i.e., $p_{ij} = p_{ji}$), then the chain is reversible.

17. Show for an ergodic chain that

(a) If the chain is reversible, then $p_{ij}p_{jk}p_{ki} = p_{ji}p_{kj}p_{ik}$.
(b) If the transition matrix has all positive entries, then the above condition assures reversibility. [HINT: Show that for fixed i the row vector $\lambda = \{p_{ij}/p_{ji}\}$ is a fixed vector of P. Hence this vector must be proportional to α.]

For the entire chapter

18. The general (finite) random walk is defined as follows. The states are numbered $s_0, s_1, \ldots, s_n$. If the process is in s_i, then it moves to s_{i-1} with probability q_i, it stays in s_i with probability r_i, and moves to s_{i+1} with probability p_i. (Where $p_i+q_i+r_i = 1$, $q_0 = 0$, $p_n = 0$.)

(a) Under what conditions is a random walk ergodic?
(b) From the equation $\alpha P = \alpha$ prove, by mathematical induction, that $a_{i+1}q_{i+1} = a_ip_i$.
(c) Prove that an ergodic random walk is reversible.
(d) Find a formula for the fixed vector α.

CHAPTER VI

FURTHER RESULTS

§ 6.1 Application of absorbing chain theory to ergodic chains. We have seen that the Z matrix enables us to find the mean and variance of the first passage time to state s_j. Assume now that we are interested in more detailed behavior of the process in going to s_j. For example, we might ask for the mean number of times that it will be in each of the other states before reaching s_j for the first time. The answer to this and other similar questions is furnished by applying the absorbing Markov chain theory. To do this we change our process by making s_j into an absorbing state. The resulting process will be an absorbing process with a single absorbing state. The behavior of this process before absorption is exactly the same as the behavior of the original process before hitting s_j for the first time. Hence we can translate all of the information we have about an absorbing Markov chain into information about our original chain. In particular it provides us with an alternative way to find the mean and variance of the first passage time from s_i to s_j, these being the mean and variance of the time before absorption in the new process. Since any proper subset of an ergodic set is an open set, we can apply the results of § 3.5 to obtain the behavior of our process before it hits a subset for the first time.

Let us illustrate the above ideas with the Land of Oz example. Assume that we are interested in the behavior of the process before the first rainy day. We make state **R** absorbing and have the new absorbing Markov chain with transition matrix:

$$P = \begin{array}{c} \\ \mathbf{R} \\ \mathbf{N} \\ \mathbf{S} \end{array} \begin{array}{c} \begin{array}{ccc} \mathbf{R} & \mathbf{N} & \mathbf{S} \end{array} \\ \begin{pmatrix} 1 & 0 & 0 \\ 1/2 & 0 & 1/2 \\ 1/4 & 1/4 & 1/2 \end{pmatrix} \end{array}.$$

The basic results for this absorbing chain are obtained from the fundamental matrix $N=(I-Q)^{-1}$ where Q is the matrix obtained by considering only non-absorbing states. For example, let n_j be the number of times the process is in state s_j before being absorbed. Then the values of $\mathbf{M}_i[\mathbf{n}_j]$ are given by the matrix N, in this case

$$N = \begin{matrix} & \mathbf{N} & \mathbf{S} \\ \mathbf{N} & 4/3 & 4/3 \\ \mathbf{S} & 2/3 & 8/3 \end{matrix} \quad \text{i.e.} \quad \begin{pmatrix} 4/3 & 4/3 \\ 2/3 & 8/3 \end{pmatrix}.$$

For example, calculated from a nice day, the mean number of nice days before the next rainy day is $4/3$. We can find $\mathbf{Var}_i[\mathbf{n}_j]$ from the matrix

$$N_2 = N(2N_{dg}-I)-N_{sq}$$

$$= \begin{matrix} & \mathbf{N} & \mathbf{S} \\ \mathbf{N} & 4/9 & 4 \\ \mathbf{S} & 2/3 & 40/9 \end{matrix} \quad \text{i.e.} \quad \begin{pmatrix} 4/9 & 4 \\ 2/3 & 40/9 \end{pmatrix}.$$

Let $\mathbf{t}$ be the function which gives the total number of steps before absorption. Then, from Theorem 3.3.5, we have that the column vector $\tau=\{\mathbf{M}_i[\mathbf{t}]\}$ is given by $\tau=N\xi$. In the example we are considering, this is

$$\tau = \begin{pmatrix} 4/3 & 4/3 \\ 2/3 & 8/3 \end{pmatrix}\begin{pmatrix} 1 \\ 1 \end{pmatrix} = \begin{pmatrix} 8/3 \\ 10/3 \end{pmatrix}.$$

The function $\mathbf{t}$ represents in the original process the time to reach state $\mathbf{R}$ for the first time. Thus the mean first passage time to $\mathbf{R}$, starting in state $\mathbf{N}$, is $8/3$, and, starting in state $\mathbf{S}$ it is $10/3$. These values agree with those found in the matrix M calculated from the Z matrix in § 4.4. Similarly, from Theorem 3.3.5, we have that the $\mathbf{Var}_i[\mathbf{t}]$ is given by the column vector $\tau_2=(2N-I)\tau-\tau_{sq}$. Calculating this, we have

$$\tau_2 = \begin{pmatrix} 56/9 \\ 62/9 \end{pmatrix}.$$

The vector τ_2 gives us the variance of the time before absorption. In terms of the original process this is the variance of the first passage time to state $\mathbf{R}$. Again the above values check with those found from the matrix M_2 obtained in § 4.5.

By successively making each state absorbing we could find all the non-diagonal elements of M and M_2 for an ergodic chain. However, the use of the Z matrix is much more natural and convenient. We would normally use the absorbing methods only to obtain the more detailed information not available by the Z matrix methods.

As an example of a cyclic chain we consider Example 2a. We make states s_1 and s_2 absorbing. We then have

$$
Q = \begin{array}{c} \\ s_3 \\ s_4 \\ s_5 \end{array}
\begin{array}{c} s_3 \quad s_4 \quad s_5 \end{array}
\begin{pmatrix} 0 & 1/2 & 0 \\ 1/2 & 0 & 1/2 \\ 0 & 1 & 0 \end{pmatrix}
\qquad
N = \begin{array}{c} \\ s_3 \\ s_4 \\ s_5 \end{array}
\begin{array}{c} s_3 \quad s_4 \quad s_5 \end{array}
\begin{pmatrix} 2 & 2 & 1 \\ 2 & 4 & 2 \\ 2 & 4 & 3 \end{pmatrix}
$$

$$
N_2 = \begin{array}{c} \\ s_3 \\ s_4 \\ s_5 \end{array}
\begin{array}{c} s_3 \quad s_4 \quad s_5 \end{array}
\begin{pmatrix} 2 & 10 & 4 \\ 2 & 12 & 6 \\ 2 & 12 & 6 \end{pmatrix}
\quad
\tau = \begin{array}{c} \\ s_3 \\ s_4 \\ s_5 \end{array}
\begin{pmatrix} 5 \\ 8 \\ 9 \end{pmatrix}
\quad
\tau_2 = \begin{array}{c} \\ s_3 \\ s_4 \\ s_5 \end{array}
\begin{pmatrix} 40 \\ 48 \\ 48 \end{pmatrix}.
$$

The entries of N and N_2 give the mean and variance of the number of times that the process is in each state before reaching s_2. (The state s_1 can only be reached through s_2.) The vectors τ and τ_2 give the mean and variance of the steps needed to reach s_2, hence of the first passage times. We can verify that the components of τ and τ_2 agree with the corresponding entries (in the second column) of M and M_2 in § 5.2.

As a second application of absorbing theory to ergodic chains, consider the following problem. Assume that we have an ergodic Markov chain with r states and that the process is observed only when it is in a subset S of the states having s elements. A new Markov chain is obtained: A single step in the new process corresponds in the old process to the transition (not necessarily in one step) from a state in S to another state in S. Let s_j and s_i be two states of S. The new transition probability will be found by finding the probability that the original process starting in s_i hits S for the first time at state s_j. This is the probability that it goes to s_j in one step, plus the probability that it goes to a state in $\tilde{S}$ and from this state enters S for the first time at state s_j. Using the results of Chapter III we can easily find these transition probabilities. To do this we relabel the states so that those in S come first. We then write the transition matrix P in the form

$$
P = \begin{array}{c} \\ S \\ \tilde{S} \end{array}
\begin{array}{c} S \qquad \tilde{S} \end{array}
\begin{pmatrix} T & U \\ R & Q \end{pmatrix}.
$$

The new process will be an s-state Markov chain with transition matrix which we denote by $\bar{P}$. We shall now find this matrix. Assume a starting state in $\tilde{S}$. Then the probability of going to each of the states in $\tilde{S}$ on the first step is given by the matrix T. To take more than one step, it must enter a state of $\tilde{\tilde{S}}$, with probabilities given by U. Then from a given state of $\tilde{\tilde{S}}$ it enters $\tilde{S}$ for the first time at state s_j with probabilities given by $(I-Q)^{-1}R$ (see Theorem 3.3.7). Putting all of this information together, we have that

$$\bar{P} = T + U(I-Q)^{-1}R.$$

It is easily seen that $\bar{P}$ again represents an ergodic chain.

6.1.1. THEOREM. *Let* $\alpha = (a_1, a_2, \ldots, a_s, a_{s+1}, \ldots, a_r)$ *be the fixed probability vector for* P. *Then* $\alpha_1 = (a_1, a_2, \ldots, a_s)$, *normalized to have sum* 1, *is the fixed probability vector for* $\bar{P}$.

PROOF. Since an ergodic chain has a unique probability vector fixed point, it is sufficient to prove that α_1 is a fixed vector for $\bar{P}$. Let $\alpha_2 = (a_{s+1}, \ldots, a_r)$. Then we can write $\alpha = (\alpha_1, \alpha_2)$. Since α is a fixed vector for P we have

$$\alpha_1 = \alpha_1 T + \alpha_2 R$$

and

$$\alpha_2 = \alpha_1 U + \alpha_2 Q.$$

From this last equation we have $\alpha_2(I-Q) = \alpha_1 U$ or

$$\alpha_2 = \alpha_1 U(I-Q)^{-1}.$$

Putting this result in the first equation we have

$$\alpha_1 = \alpha_1 T + \alpha_1 U(I-Q)^{-1}R$$

which states that α_1 is a fixed vector for $\bar{P}$.

As an example of the above procedure let us consider Example 6. The transition matrix for this random walk example is

$$P = \begin{array}{c} \\ s_1 \\ s_2 \\ s_3 \\ s_4 \\ s_5 \end{array} \begin{array}{c} \begin{array}{ccccc} s_1 & s_2 & s_3 & s_4 & s_5 \end{array} \\ \left(\begin{array}{cc|ccc} 0 & 1/4 & 1/4 & 1/4 & 1/4 \\ 1/3 & 1/3 & 1/3 & 0 & 0 \\ \hline 0 & 1/3 & 1/3 & 1/3 & 0 \\ 0 & 0 & 1/3 & 1/3 & 1/3 \\ 1/4 & 1/4 & 1/4 & 1/4 & 0 \end{array}\right) \end{array}.$$

The fixed vector for this Markov chain is $\alpha = (4/38, 9/38, 12/38, 9/38, 4/38)$. Assume now that the process is observed only when it is in s_1 and s_2.

Then we find the new transition matrix as follows. From the discussion of this example in 3.5.5 we have

$$(I-Q)^{-1} = \begin{pmatrix} 21/9 & 12/9 & 4/9 \\ 15/9 & 24/9 & 8/9 \\ 9/9 & 9/9 & 12/9 \end{pmatrix}.$$

Thus the new transition probabilities are

$$\begin{aligned}\bar{P} &= T+U(I-Q)^{-1}R \\ &= \begin{pmatrix} 0 & 1/4 \\ 1/3 & 1/3 \end{pmatrix} + \begin{pmatrix} 1/4 & 1/4 & 1/4 \\ 1/3 & 0 & 0 \end{pmatrix}\begin{pmatrix} 21/9 & 12/9 & 4/9 \\ 15/9 & 24/9 & 8/9 \\ 9/9 & 9/9 & 12/9 \end{pmatrix}\begin{pmatrix} 0 & 1/3 \\ 0 & 0 \\ 1/4 & 1/4 \end{pmatrix} \\ &= \begin{pmatrix} 1/6 & 5/6 \\ 10/27 & 17/27 \end{pmatrix}.\end{aligned}$$

The fixed vector for $\bar{P}$ is $\bar{\alpha} = (4/13, 9/13)$ which is simply the first two components of α normalized to have sum 1.

For a cyclic example, let us consider the random walk Example 2a. We observe the process in $\mathbf{S} = \{s_1, s_2, s_3\}$.

$$P = \left(\begin{array}{ccc|cc} 0 & 1 & 0 & 0 & 0 \\ 1/2 & 0 & 1/2 & 0 & 0 \\ 0 & 1/2 & 0 & 1/2 & 0 \\ \hline 0 & 0 & 1/2 & 0 & 1/2 \\ 0 & 0 & 0 & 1 & 0 \end{array}\right); \quad \alpha = (1/8, 1/4, 1/4, 1/4, 1/8).$$

$$\bar{P} = \begin{pmatrix} 0 & 1 & 0 \\ 1/2 & 0 & 1/2 \\ 0 & 1/2 & 0 \end{pmatrix} + \begin{pmatrix} 0 & 0 \\ 0 & 0 \\ 1/2 & 0 \end{pmatrix}\begin{pmatrix} 1 & -1/2 \\ -1 & 1 \end{pmatrix}^{-1}\begin{pmatrix} 0 & 0 & 1/2 \\ 0 & 0 & 0 \end{pmatrix}.$$

$$\bar{P} = \begin{pmatrix} 0 & 1 & 0 \\ 1/2 & 0 & 1/2 \\ 0 & 1/2 & 1/2 \end{pmatrix}; \quad \bar{\alpha} = (1/5, 2/5, 2/5).$$

$\bar{P}$ differs only slightly from the first three states of P. The process can leave $\mathbf{S}$ only through s_3, and must return there. Hence only p_{33} is changed. We note that, in accordance with § 6.1.1, $\bar{\alpha}$ consists of the first three components of α normalized.

§ 6.2 Application of ergodic chain theory to absorbing Markov chains. In the preceding section we saw that absorbing Markov chain theory could furnish us with new information about ergodic chains. We shall now show that certain results of absorbing chain theory can be obtained by using the theory of ergodic chains.

We will need the following generalization of § 5.1.2(b).

6.2.1 THEOREM. *Every Markov chain with a single ergodic set has a unique probability vector fixed point. This vector has positive components for the ergodic states, and zero for the transient states.*

PROOF. Let us write the transition matrix in canonical form.

$$P = \left(\begin{array}{c|c} S & O \\ \hline R & Q \end{array}\right).$$

The matrix S is the transition matrix of the ergodic set; hence it has a limiting vector $\alpha_1 > 0$. Let $\alpha = (\alpha_1, \alpha_2)$, where α_2 is a vector with s components all 0. Then we see that α is a probability vector fixed point of P. Conversely, let us suppose that $\beta = (\beta_1, \beta_2)$ is a probability vector fixed point. Then $\beta_2 Q = \beta_2$. Hence $\beta_2 Q^n = \beta_2$, and $\beta_2 = \lim_{n\to\infty} \beta_2 Q^n = 0$. Thus β_1 is a probability vector fixed point of S; hence by § 5.1.2(b) we have $\beta_1 = \alpha_1$, and $\beta = \alpha$.

Assume now that we have an absorbing Markov chain with r states, $r - s$ of which are absorbing, and s non-absorbing. As usual we shall label the absorbing states so that they come first. The transition matrix then has the form:

$$P = \left(\begin{array}{c|c} I & O \\ \hline R & Q \end{array}\right).$$

We now change this process into a new process as follows. Let $\pi = \{p_1, p_2, \ldots, p_r\}$ be the initial probability vector for the given process. Whenever this process reaches an absorbing state it is started over again with the same initial vector π. The resulting process is a new Markov chain with transition matrix given by

$$P' = \left(\begin{array}{c|c} p_1, p_2, \ldots, p_{r-s} & p_{r-s+1}, \ldots, p_r \\ p_1, p_2, \ldots, p_{r-s} & p_{r-s+1}, \ldots, p_r \\ p_1, p_2, \ldots, p_{r-s} & p_{r-s+1}, \ldots, p_r \\ \hline R & Q \end{array}\right).$$

The matrix P' is obtained by making all rows corresponding to absorbing states the same vector π. Let $\pi_1 = (p_1, \ldots, p_{r-s})$ and $\pi_2 = (p_{r-s+1}, \ldots, p_r)$. Then P' may be written in the form:

$$P' = \left(\begin{array}{c|c} \xi_{r-s}\pi_1 & \xi_{r-s}\pi_2 \\ \hline R & Q \end{array}\right).$$

6.2.2 THEOREM. *The matrix P' represents a Markov chain with a single ergodic set.*

PROOF. Let $\mathbf{I}$ be the set of states for which π has positive components. Let $\mathbf{J}$ be the set of all states to which the process can go starting in $\mathbf{I}$. It is clear that from s_a, $a = 1, 2, \ldots, r-s$, we can go only to states in $\mathbf{J}$. However, from any state we can go to some s_a, since the original chain was absorbing. Hence all states in $\tilde{\mathbf{J}}$ are transient.

Since from any state we can go to an s_a, and from this to all states in $\mathbf{I}$, and hence in $\mathbf{J}$, we see that $\mathbf{J}$ is an ergodic set. Hence the new process has the single ergodic set $\mathbf{J}$, and at least one $s_a \in \mathbf{J}$.

Let α be the fixed probability vector for P'. Write α in the form $\alpha = (\alpha_1, \alpha_2)$ where $\alpha_1 = (a_1, a_2, \ldots, a_{r-s})$ and $\alpha_2 = (a_{r-s+1}, a_{r-s+2}, \ldots, a_r)$. Then, since $\alpha P' = \alpha$, we have the two equations

$$\alpha_1\xi_{r-s}\pi_1 + \alpha_2 R = \alpha_1 \tag{1}$$

$$\alpha_1\xi_{r-s}\pi_2 + \alpha_2 Q = \alpha_2. \tag{2}$$

By § 6.2.1 we know that $\alpha_1 > 0$, hence $\alpha_1\xi_{r-s} > 0$. Let $\bar{\alpha} = \dfrac{1}{\alpha_1\xi_{r-s}}\,\alpha$. The result $\bar{\alpha} = \{\bar{\alpha}_1, \bar{\alpha}_2\}$ will still be a fixed vector; remembering that $\bar{\alpha}_1\xi_{r-s} = 1$, our equations become

$$\pi_1 + \bar{\alpha}_2 R = \bar{\alpha}_1 \tag{1'}$$

$$\pi_2 + \bar{\alpha}_2 Q = \bar{\alpha}_2. \tag{2'}$$

From equation (2′) we have

$$\bar{\alpha}_2 = \pi_2(I-Q)^{-1}.$$

This inverse exists because the original chain was absorbing. From Theorem 3.3.5 we see that $\bar{\alpha}_2$ gives us the mean number of times in each of the states before absorption, for the given initial probability vector π.

Using the result just obtained in equation (1′) we have

$$\bar{\alpha}_1 = \pi_1 + \pi_2(I-Q)^{-1}R.$$

The vector π_1 gives the probability in the original process of being absorbed at each absorbing state on the initial step, and $\pi_2(I-Q)^{-1}R$ gives this probability for being absorbed in each absorbing state if the initial step is to a non-absorbing state. Hence $\bar{\alpha}_1$ gives the probabilities for absorption in each of the given states, for the initial probability vector π.

We thus see that the single vector $\bar{\alpha}$ furnishes us with both absorption probabilities and the mean number of times in a transient state before absorption. This method is more economical than the method of Chapter III, if we are interested in a given initial probability vector. It must be remembered, however, that Chapter III furnishes the solution for any initial vector.

Let us carry out this procedure for the random walk Example 1. The transition matrix is

$$P = \begin{array}{c} \\ s_1 \\ s_5 \\ s_2 \\ s_3 \\ s_4 \end{array} \begin{array}{c} \begin{array}{ccccc} s_1 & s_5 & s_2 & s_3 & s_4 \end{array} \\ \begin{pmatrix} 1 & 0 & 0 & 0 & 0 \\ 0 & 1 & 0 & 0 & 0 \\ q & 0 & 0 & p & 0 \\ 0 & 0 & q & 0 & p \\ 0 & p & 0 & q & 0 \end{pmatrix} \end{array}.$$

Let $\pi = (0, 0, 0, 1, 0)$. Then the new Markov chain obtained by the above procedure is

$$P' = \begin{array}{c} \\ s_1 \\ s_5 \\ s_2 \\ s_3 \\ s_4 \end{array} \begin{array}{c} \begin{array}{ccccc} s_1 & s_5 & s_2 & s_3 & s_4 \end{array} \\ \begin{pmatrix} 0 & 0 & 0 & 1 & 0 \\ 0 & 0 & 0 & 1 & 0 \\ q & 0 & 0 & p & 0 \\ 0 & 0 & q & 0 & p \\ 0 & p & 0 & q & 0 \end{pmatrix} \end{array}.$$

This is the same as Example 3 of § 2.2, with the states reordered. The fixed vector is

$$\alpha = \frac{1}{2+p^2+q^2}(q^2, p^2, q, 1, p).$$

Thus

$$\bar{\alpha} = \frac{1}{p^2+q^2}(q^2, p^2, q, 1, p).$$

From this we see that, in the original process, the absorption probabilities are $q^2/(p^2+q^2)$ for state s_1 and $p^2/(p^2+q^2)$ for state s_5. The mean number of times in each of the states s_2, s_3, s_4 are $q/(p^2+q^2)$, $1/(p^2+q^2)$, and $p/(p^2+q^2)$ respectively. This is in agreement with the results found in § 3.4.1.

If we let $\pi=(0, 0, 0, 0, 1)$, then the resulting chain is cyclic with $d=2$. The same is true if $\pi=(0, 0, c, 0, d)$, $d=1-c$. We will work out this example.

$$P' = \begin{matrix} s_1 \\ s_5 \\ s_2 \\ s_3 \\ s_4 \end{matrix} \begin{pmatrix} 0 & 0 & c & 0 & d \\ 0 & 0 & c & 0 & d \\ q & 0 & 0 & p & 0 \\ 0 & 0 & q & 0 & p \\ 0 & p & 0 & q & 0 \end{pmatrix}.$$

In calculating $\bar{\alpha}$ it is simplest *not* to find α first. In solving the equation $\bar{\alpha}P'=\bar{\alpha}$, the condition $\bar{a}_1+\bar{a}_2=1$ is very helpful.

$$\bar{\alpha} = \frac{1}{p^2+q^2}(q^2+p^2qc-pq^2d,\ p^2+pq^2d-p^2qc,$$
$$q+p^2c-pqd,\ 1-qc-pd,\ p+q^2d-pqc).$$

If we let $p=q={}^1/_2$, we obtain

$$\bar{\alpha} = ({}^1/_2+{}^1/_4(c-d),\ {}^1/_2-{}^1/_4(c-d),\ 1+{}^1/_2(c-d),\ 1,\ 1-{}^1/_2(c-d)).$$

The first two components furnish the probabilities of absorption in s_1, s_5 for the chain starting with π. As is to be expected, the larger c is, the more likely it is that the process is absorbed in s_1. The last three components furnish the mean number of times in a state before absorption. It is interesting to note that for s_3 this is 1, no matter what c is.

An interesting application of the last result can be made to ergodic chain theory. Let P be the transition matrix for an ergodic chain with fixed probability vector α. Let us make one of the states, say s_1, into an absorbing state. Then every time this process reaches s_1 we will start it again with the probabilities $\pi=\{p_{1j}\}$. Then by the above result the fixed vector for this new process will give us the mean number of times in each state before absorption. But the new process is just the original process, and time before absorption means time between occurrences of state s_1. Hence by re-normalizing α to have first component 1 we will obtain the mean number of times in each state between occurrences of state s_1 for the original ergodic chain. Since s_1 was arbitrary, this gives us the following theorem:

6.2.3 THEOREM. *Let α be the fixed probability vector for an ergodic* *chain. Then the mean number of times in state* s_j *between occurrences of state* s_i *is* a_j/a_i.

Note that if the transition matrix for the chain has column sums 1, then the fixed vector has all components equal. This means, by this theorem, that the mean number of times in each of the other states, between occurrences of a given state, is the same.

6.2.4 COROLLARY. *Let* $\bar{\alpha}$ *be the vector obtained from* α *by deleting component* l; *let* ρ *be the* l*-th row of* P *with component* l *deleted; let* Q *be the matrix obtained from* P *by deleting row* l *and column* l; *and let* $N=(I-Q)^{-1}$, $\tau=N\xi$. *Then*

$$\frac{1}{a_l}\bar{\alpha} = \rho N, \tag{a}$$

$$\frac{1}{a_l} = 1+\rho\tau \tag{b}$$

PROOF. In (a) the left side is the mean number of times in each of the other states between occurrences of s_l. The right side is the same quantity computed from absorbing chain theory. In (b) we have m_{ll} computed from regular and absorbing chain theory, respectively.

6.2.5 THEOREM. *If* Z *is the fundamental matrix of an ergodic chain, and* A *is its limiting matrix, and* N *is the fundamental matrix of the absorbing chain obtained by making* s_l *absorbing, and we construct* N^* *from* N *by inserting an* l*-th row and* l*-th column of all zeros, then*

$$Z = A+(I-A)N^*(I-A). \tag{3}$$

PROOF. Without loss of generality we may choose $i=1$. Then, using the notation of § 6.2.4, P and A are of the form

$$P = \left(\begin{array}{c|c} p_{11} & \rho \\ \hline \xi - Q\xi & Q \end{array}\right) \qquad A = \left(\begin{array}{c|c} a_1 & \bar{\alpha} \\ \hline a_1\xi & \xi\bar{\alpha} \end{array}\right),$$

and hence,

$$I-P = \left(\begin{array}{c|c} 1-p_{11} & -\rho \\ \hline Q\xi-\xi & I-Q \end{array}\right) \qquad I-A = \left(\begin{array}{c|c} 1-a_1 & -\bar{\alpha} \\ \hline -a_1\xi & I-\xi\bar{\alpha} \end{array}\right),$$

and

$$N^* = \left(\begin{array}{c|c} 0 & O \\ \hline O & N \end{array}\right).$$

Then

$$(I-P)N^* = \left(\begin{array}{c|c} 0 & -\rho N \\ \hline 0 & I \end{array}\right)$$

$$(I-P)N^*\,(I-A) = \begin{pmatrix} a_1\rho\tau & -\rho N+\rho\tau\bar{\alpha} \\ -a_1\xi & I-\xi\bar{\alpha} \end{pmatrix}.$$

Making use of § 6.2.4,

$$\begin{aligned} (I-P)N^*\,(I-A) &= I-A \\ [I-P+A][A+(I-A)N^*(I-A)] &= A+(I-P)N^*(I-A) \\ &= A+(I-A) \\ &= I. \end{aligned}$$

Hence

$$A+(I-A)N^*(I-A) = (I-P+A)^{-1} = Z.$$

It is interesting to note that in (3) we may use any N^* obtained by making any one state absorbing.

6.2.6 COROLLARY. *If in § 6.2.5 we let* $N=\{n^{(l)}{}_{ij}\}$ *and* $N\xi=\{t^{(l)}{}_i\}$, *and* $n^{(l)}{}_{ij}=n^{(l)}{}_{ji}=t^{(l)}{}_i=0$ *if* $i=l$, *then*

$$z_{ij} = a_j+n^{(l)}{}_{ij}-\sum_{k\neq l} a_k n^{(l)}{}_{kj}-a_j t^{(l)}{}_i+a_j\sum_{k\neq l} a_k t^{(l)}{}_k. \tag{a}$$

$$m_{ij} = (1/a_j)(n^{(l)}{}_{jj}-n^{(l)}{}_{ij}+d_{ij})+t^{(l)}{}_i-t^{(l)}{}_j. \tag{b}$$

These quantities are obtained directly from § 6.2.5, making use of § 4.4.7 for (b). These formulas may be used to derive many interesting results. A few of these are given below. The number $n^{(l)}{}_{ij}$ is the number of times the process is in s_j, starting in s_i, before reaching s_l for the first time.

6.2.7 COROLLARY.

(a) $m_{ij}+m_{jl} = m_{il}+\dfrac{n^{(l)}{}_{jj}}{a_j}(1-h^{(l)}{}_{ij}),$ *for* $i, j \neq l$.

(b) $m_{jl}+m_{lj} = \dfrac{n^{(l)}{}_{jj}}{a_j}.$ (c) $\dfrac{n^{(l)}{}_{jj}}{n^{(j)}{}_{ll}} = \dfrac{a_j}{a_l}.$

PROOF. From § 6.2.6(b), if $i, j\neq l$,

$$m_{ij}+m_{jl} = m_{ij}+t^{(l)}{}_j = (1/a_j)(n^{(l)}{}_{jj}-n^{(l)}{}_{ij}+d_{ij})+m_{il}$$

$$1-h^{(l)}{}_{ij} = 1-\frac{n^{(l)}{}_{ij}-d_{ij}}{n^{(l)}{}_{jj}} = \frac{n^{(l)}{}_{jj}-n^{(l)}{}_{ij}+d_{ij}}{n^{(l)}{}_{jj}}.$$

Hence (a) follows.

$$m_{jl}+m_{lj} = t^{(l)}{}_j+(1/a_j)n^{(l)}{}_{jj}-t^{(l)}{}_j.$$

Hence (b) follows.

$$1 = \frac{m_{jl}+m_{lj}}{m_{lj}+m_{jl}} = \frac{n^{(l)}{}_{jj}/a_j}{n^{(j)}{}_{ll}/a_l}.$$

Hence (c) follows.

If in the Land of Oz example we make **R** absorbing, we obtain (see § 6.1),

$$N = \begin{pmatrix} 4/3 & 4/3 \\ 2/3 & 8/3 \end{pmatrix}.$$

$$\begin{aligned} A+(I-A)N^*(I-A) \\ &= \begin{pmatrix} 2/5 & 1/5 & 2/5 \\ 2/5 & 1/5 & 2/5 \\ 2/5 & 1/5 & 2/5 \end{pmatrix} \\ &+ \begin{pmatrix} 3/5 & -1/5 & -2/5 \\ -2/5 & 4/5 & -2/5 \\ -2/5 & -1/5 & 3/5 \end{pmatrix} \begin{pmatrix} 0 & 0 & 0 \\ 0 & 4/3 & 4/3 \\ 0 & 2/3 & 8/3 \end{pmatrix} \begin{pmatrix} 3/5 & -1/5 & -2/5 \\ -2/5 & 4/5 & -2/5 \\ -2/5 & -1/5 & 3/5 \end{pmatrix} \\ &= 1/75 \begin{pmatrix} 86 & 3 & -14 \\ 6 & 63 & 6 \\ -14 & 3 & 86 \end{pmatrix} = Z \end{aligned}$$

as we saw in § 4.3. Using results from § 4.4 and from § 6.1, we can illustrate Corollaries 6.2.6 and 6.2.7.

$$\begin{aligned} m_{NS} &= (1/a_S)(n_{SS}-n_{NS})+t_N-t_S \\ &= (5/2)(8/3-4/3)+8/3-10/3 = 8/3 \end{aligned}$$

$$m_{SR}+m_{RS} = \frac{n_{SS}}{a_S} \quad \text{or} \quad 10/3+10/3 = (8/3)/(2/5).$$

§ 6.3 Combining states. Assume that we are given an r-state Markov chain with transition matrix P and initial vector π. Let $\mathbf{A} = \{\mathbf{A}_1, \mathbf{A}_2, \ldots, \mathbf{A}_t\}$ be a partition of the set of states. We form a new process as follows. The outcome of the j-th experiment in the new process is the set $\mathbf{A}_k$ that contains the outcome of the j-th step in the original chain. We define branch probabilities as follows: At the zero level we assign

$$\mathbf{Pr}_\pi[\mathbf{f}_0 \in \mathbf{A}_i]. \tag{1}$$

At the first level we assign

$$\mathbf{Pr}_\pi[\mathbf{f}_1 \in \mathbf{A}_j | \mathbf{f}_0 \in \mathbf{A}_i].$$

In general, at the n-th level we assign branch probabilities,

$$\mathbf{Pr}_\pi[\mathbf{f}_n \in \mathbf{A}_t | \mathbf{f}_{n-1} \in \mathbf{A}_s \wedge \cdots \wedge \mathbf{f}_1 \in \mathbf{A}_j \wedge \mathbf{f}_0 \in \mathbf{A}_i]. \tag{2}$$

The above procedure could be used to reduce a process with a very large number of states to a process with a smaller number of states. We call this process a *lumped process.* It is also often the case in applications that we are only interested in questions which relate to this coarser analysis of the possibilities. Thus it is important to be able to determine whether the new process can be treated by Markov chain methods.

6.3.1 DEFINITION. *We shall say that a Markov chain is* lumpable *with respect to a partition* $\mathbf{A} = \{\mathbf{A}_1, \mathbf{A}_2, \ldots, \mathbf{A}_r\}$ *if for every starting vector* π *the lumped process defined by* (1) *and* (2) *is a Markov chain and the transition probabilities do not depend on the choice of* π.

We shall see in the next section that, at least for regular chains, the condition that the transition probabilities do not depend on π follows from the requirement that every starting vector give a Markov chain.

Let $p_{iA_j} = \sum_{\mathbf{s}_k \in \mathbf{A}_j} p_{ik}$. Then p_{iA_j} represents the probability of moving from state $\mathbf{s}_i$ into set $\mathbf{A}_j$ in one step for the original Markov chain.

6.3.2 THEOREM. *A necessary and sufficient condition for a Markov chain to be lumpable with respect to a partition* $\mathbf{A} = \{\mathbf{A}_1, \mathbf{A}_2, \ldots, \mathbf{A}_s\}$ *is that for every pair of sets* $\mathbf{A}_i$ *and* $\mathbf{A}_j$, p_{kA_j} *have the same value for every* $\mathbf{s}_k$ *in* $\mathbf{A}_i$. *These common values* $\{\hat{p}_{ij}\}$ *form the transition matrix for the lumped chain.*

PROOF. For the chain to be lumpable it is clearly necessary that

$$\mathbf{Pr}_\pi[\mathbf{f}_1 \in \mathbf{A}_j | \mathbf{f}_0 \in \mathbf{A}_i]$$

be the same for every π for which it is defined. Call this common value $\hat{p}_{ij}$. In particular this must be the same for π having a 1 in its k-th component, for state $\mathbf{s}_k$ in $\mathbf{A}_i$. Hence $p_{kA_j} = \mathbf{Pr}_k[\mathbf{f}_1 \in \mathbf{A}_j] = \hat{p}_{ij}$ for every $\mathbf{s}_k$ in $\mathbf{A}_i$. Thus the condition given is necessary. To prove it is sufficient, we must show that if the condition is satisfied the probability (2) depends only on $\mathbf{A}_s$ and $\mathbf{A}_t$. The probability (2) may be written in the form

$$\mathbf{Pr}_{\pi'}[\mathbf{f}_1 \in \mathbf{A}_t]$$

where π' is a vector with non-zero components only on the states of $\mathbf{A}_s$. It depends on π and on the first n outcomes. However, if $\mathbf{Pr}_k[\mathbf{f}_1 \in \mathbf{A}_t] = \hat{p}_{st}$ for all $\mathbf{s}_k$ in $\mathbf{A}_s$, then it is clear also that $\mathbf{Pr}_{\pi'}[\mathbf{f}_1 \in \mathbf{A}_t] = \hat{p}_{st}$. Thus the probability in (2) depends only on $\mathbf{A}_s$ and $\mathbf{A}_t$.

6.3.3 EXAMPLE. Let us consider the Land of Oz example. Recall that P is given by

$$P = \begin{array}{c} \\ \mathbf{R} \\ \mathbf{N} \\ \mathbf{S} \end{array} \begin{array}{c} \begin{array}{ccc} \mathbf{R} & \mathbf{N} & \mathbf{S} \end{array} \\ \begin{pmatrix} 1/2 & 1/4 & 1/4 \\ 1/2 & 0 & 1/2 \\ 1/4 & 1/4 & 1/2 \end{pmatrix} \end{array}.$$

Assume now that we are interested only in "good" and "bad" weather. This suggests lumping **R** and **S**. We note that the probability of moving from either of these states to **N** is the same. Hence if we choose for our partition $\mathbf{A} = (\{\mathbf{N}\}, \{\mathbf{R},\mathbf{S}\}) = (\mathbf{G}, \mathbf{B})$, the condition for lumpability is satisfied. The new transition matrix is

$$P = \begin{array}{c} \\ \mathbf{G} \\ \mathbf{B} \end{array} \begin{array}{c} \begin{array}{cc} \mathbf{G} & \mathbf{B} \end{array} \\ \begin{pmatrix} 0 & 1 \\ 1/4 & 3/4 \end{pmatrix} \end{array}.$$

Note that the condition for lumpability is not satisfied for the partition $\mathbf{A} = (\{\mathbf{R}\}, \{\mathbf{N},\mathbf{S}\})$ since $p_{NA_1} = p_{NR} = 1/2$ and $p_{SA_1} = p_{SR} = 1/4$.

Assume now that we have a Markov chain which is lumpable with respect to a partition $\mathbf{A} = \{\mathbf{A}_1, \ldots, \mathbf{A}_s\}$. We assume that the original chain had r states and the lumped chain has s states. Let U be the $s \times r$ matrix whose i-th row is the probability vector having equal components for states in $\mathbf{A}_i$ and 0 for the remaining states. Let V be the $r \times s$ matrix with the j-th column a vector with 1's in the components corresponding to states in $\mathbf{A}_j$ and 0's otherwise. Then the lumped transition matrix is given by

$$\hat{P} = UPV.$$

In the Land of Oz example this is

$$\hat{P} = \overset{U}{\begin{pmatrix} 0 & 1 & 0 \\ 1/2 & 0 & 1/2 \end{pmatrix}} \overset{P}{\begin{pmatrix} 1/2 & 1/4 & 1/4 \\ 1/2 & 0 & 1/2 \\ 1/4 & 1/4 & 1/2 \end{pmatrix}} \overset{V}{\begin{pmatrix} 0 & 1 \\ 1 & 0 \\ 0 & 1 \end{pmatrix}}$$

$$= \overset{U}{\begin{pmatrix} 0 & 1 & 0 \\ 1/2 & 0 & 1/2 \end{pmatrix}} \overset{PV}{\begin{pmatrix} 1/4 & 3/4 \\ 0 & 1 \\ 1/4 & 3/4 \end{pmatrix}} = \begin{pmatrix} 0 & 1 \\ 1/4 & 3/4 \end{pmatrix}.$$

Note that the rows of PV corresponding to the elements in the same set of the partition are the same. This will be true in general for a chain which satisfies the condition for lumpability. The matrix U then simply removes this duplication of rows. The choice of U is by no means unique. In fact, all that is needed is that the i-th row should be a probability vector with non-zero components only for states in $\mathbf{A}_i$. We have chosen, for convenience, the vector with equal components for these states. Also it is convenient for proofs to assume that the states are numbered so that those in $\mathbf{A}_1$ come first, those in $\mathbf{A}_2$ come next, etc. In all proofs we shall understand that this had been done.

The following result will be useful in deriving formulas for lumped chains.

6.3.4 THEOREM. *If P is the transition matrix of a chain lumpable with respect to the partition* $\mathbf{A}$, *and if the matrices U and V are defined —as above—with respect to this partition, then*

$$VUPV = PV. \tag{3}$$

PROOF. The matrix VU has the form

$$VU = \left(\begin{array}{c|c|c} W_1 & 0 & 0 \\ \hline 0 & W_2 & 0 \\ \hline 0 & 0 & W_3 \end{array}\right),$$

where W_1, W_2, and W_3 are probability matrices. Condition (3) states that the columns of PV are fixed vectors of VU. But since the chain is lumpable, the probability of moving from a state of $\mathbf{A}_i$ to the set $\mathbf{A}_j$ is the same for all states in $\mathbf{A}_i$, hence the components of a column of PV corresponding to $\mathbf{A}_j$ are all the same. Therefore they form a fixed vector for W_j. This proves (3).

6.3.5 THEOREM. *If P,* $\mathbf{A}$, *U, and V are as in Theorem* 6.3.4, *then condition* (3) *is equivalent to lumpability.*

PROOF. We have already seen that (3) is implied by lumpability. Conversely, let us suppose that (3) holds. Then the columns of PV are fixed vectors for VU. But each W_j is the transition matrix of an ergodic chain, hence its only fixed column vectors are of the form $c\xi$. Hence all the components of a column of PV corresponding to one set $\mathbf{A}_j$ must be the same. That is, the chain is lumpable by § 6.3.2.

Note that from (3)

$$\hat{P}^2 = UPVUPV$$
$$= UP^2V$$

and in general

$$\hat{P}^n = UP^nV.$$

This last fact could also be verified directly from the process.

Assume now that P is an absorbing chain. We shall restrict our discussion to the case where we lump only states of the same kind. That is, any subset of our partition will contain only absorbing states or only non-absorbing states. We recall that the standard form for an absorbing chain is

$$P = \left(\begin{array}{c|c} I & O \\ \hline R & Q \end{array}\right).$$

We shall write U in the form

$$U = \left(\begin{array}{c|c} U_1 & O \\ \hline O & U_2 \end{array}\right),$$

where entries of U_1 refer to absorbing states and entries of U_2 to non-absorbing states. Similarly we write V in the form

$$V = \left(\begin{array}{c|c} V_1 & O \\ \hline O & V_2 \end{array}\right).$$

Then, if we consider the condition for lumpability, $VUPV = PV$, we obtain in terms of the above matrices the equivalent set of conditions:

$$V_1U_1V_1 = V_1 \tag{4a}$$

$$V_2U_2RV_1 = RV_1 \tag{4b}$$

$$V_2U_2QV_2 = QV_2. \tag{4c}$$

Since $U_1V_1 = I$, the first condition is automatically satisfied.

The standard form for the transition matrix $\hat{P}$ is obtained from

$$\hat{P} = UPV$$

$$\hat{P} = \left(\begin{array}{c|c} U_1 & O \\ \hline O & U_2 \end{array}\right)\left(\begin{array}{c|c} I & O \\ \hline R & Q \end{array}\right)\left(\begin{array}{c|c} V_1 & O \\ \hline O & V_2 \end{array}\right).$$

Multiplying this out we obtain

$$\hat{P} = \left(\begin{array}{c|c} I & 0 \\ \hline U_2RV_1 & U_2QV_2 \end{array}\right).$$

Hence we have

$$\hat{R} = U_2RV_1$$
$$\hat{Q} = U_2QV_2.$$

From condition (4c) we obtain

$$\hat{Q}^2 = U_2QV_2U_2QV_2$$
$$= U_2Q^2V_2.$$

More generally we have

$$\hat{Q}^n = U_2Q^nV_2.$$

From the infinite series representation for the fundamental matrix N we have

$$\hat{N} = I+\hat{Q}+\hat{Q}^2+\cdots$$
$$= U_2IV_2+U_2QV_2+\cdots$$
$$= U_2(I+Q+Q^2+\cdots)V_2$$
$$\hat{N} = U_2NV_2.$$

From this we obtain

$$\hat{\tau} = U_2NV_2\xi$$
$$\hat{\tau} = U_2N\xi$$
$$\hat{\tau} = U_2\tau$$

and

$$\hat{B} = \hat{N}\hat{R} = U_2NV_2U_2RV_1$$
$$\hat{B} = U_2NRV_1$$
$$\hat{B} = U_2BV_1.$$

Hence all three of the quantities N, τ, and B are easily obtained for the lumped chain from the corresponding quantities for the original chain.

An important consequence of our result $\hat{\tau} = U_2\tau$ is the following. Let $\mathbf{A}_i$ be any non-absorbing set, and $\mathbf{s}_k$ be a state in $\mathbf{A}_i$. We can choose the i-th row of U_2 to be a probability vector with 1 in the $\mathbf{s}_k$ component. But this means that $t_i = t_k$ for all $\mathbf{s}_k$ in $\mathbf{A}_i$. Hence when a chain is lumpable, the mean time to absorption must be the same for all starting states $\mathbf{s}_k$ in the same set $\mathbf{A}_i$

As an example of the above, let us consider the random walk example with transition matrix

$$P = \begin{array}{c} \\ s_1 \\ s_5 \\ s_2 \\ s_3 \\ s_4 \end{array} \begin{array}{c} \begin{array}{ccccc} s_1 & s_5 & s_2 & s_3 & s_4 \end{array} \\ \left(\begin{array}{cc|ccc} 1 & 0 & 0 & 0 & 0 \\ 0 & 1 & 0 & 0 & 0 \\ \hline 1/2 & 0 & 0 & 1/2 & 0 \\ 0 & 0 & 1/2 & 0 & 1/2 \\ 0 & 1/2 & 0 & 1/2 & 0 \end{array}\right) \end{array}.$$

We take the partition $\mathbf{A} = (\{s_1, s_5\}, \{s_2, s_4\}, \{s_3\})$. For this partition the condition for lumpability is satisfied. Notice that this would not have been the case if we have unequal probabilities for moving to the right or left.

From the original chain we found

$$N = \begin{array}{c} \\ s_2 \\ s_3 \\ s_4 \end{array} \begin{array}{c} \begin{array}{ccc} s_2 & s_3 & s_4 \end{array} \\ \begin{pmatrix} 3/2 & 1 & 1/2 \\ 1 & 2 & 1 \\ 1/2 & 1 & 3/2 \end{pmatrix} \end{array}$$

$$\tau = \begin{pmatrix} 3 \\ 4 \\ 3 \end{pmatrix}$$

$$B = \begin{array}{c} \\ s_2 \\ s_3 \\ s_4 \end{array} \begin{array}{c} \begin{array}{cc} s_1 & s_5 \end{array} \\ \begin{pmatrix} 3/4 & 1/4 \\ 1/2 & 1/2 \\ 1/4 & 3/4 \end{pmatrix} \end{array}.$$

The corresponding quantities for the lumped process are

$$\hat{P} = \begin{pmatrix} 1/2 & 1/2 & 0 & 0 & 0 \\ 0 & 0 & 1/2 & 0 & 1/2 \\ 0 & 0 & 0 & 1 & 0 \end{pmatrix} \begin{pmatrix} 1 & 0 & 0 & 0 & 0 \\ 0 & 1 & 0 & 0 & 0 \\ 1/2 & 0 & 0 & 1/2 & 0 \\ 0 & 0 & 1/2 & 0 & 1/2 \\ 0 & 1/2 & 0 & 1/2 & 0 \end{pmatrix} \begin{pmatrix} 1 & 0 & 0 \\ 1 & 0 & 0 \\ 0 & 1 & 0 \\ 0 & 0 & 1 \\ 0 & 1 & 0 \end{pmatrix}$$

$$= \begin{array}{c} \\ A_1 \\ A_2 \\ A_3 \end{array} \begin{array}{c} \begin{array}{ccc} A_1 & A_2 & A_3 \end{array} \\ \begin{pmatrix} 1 & 0 & 0 \\ 1/2 & 0 & 1/2 \\ 0 & 1 & 0 \end{pmatrix} \end{array}$$

$$\hat{N} = \begin{pmatrix} 1/2 & 0 & 1/2 \\ 0 & 1 & 0 \end{pmatrix} \begin{pmatrix} 3/2 & 1 & 1/2 \\ 1 & 2 & 1 \\ 1/2 & 1 & 3/2 \end{pmatrix} \begin{pmatrix} 1 & 0 \\ 0 & 1 \\ 1 & 0 \end{pmatrix}$$

$$= \begin{matrix} & A_2 & A_3 \\ A_2 & 2 & 1 \\ A_3 & 2 & 2 \end{matrix}$$

$$\hat{\tau} = \begin{pmatrix} 1/2 & 0 & 1/2 \\ 0 & 1 & 0 \end{pmatrix} \begin{pmatrix} 3 \\ 4 \\ 3 \end{pmatrix} = \begin{matrix} A_2 \\ A_3 \end{matrix} \begin{pmatrix} 3 \\ 4 \end{pmatrix}$$

$$\hat{B} = \begin{pmatrix} 1/2 & 0 & 1/2 \\ 0 & 1 & 0 \end{pmatrix} \begin{pmatrix} 3/4 & 1/4 \\ 1/2 & 1/2 \\ 1/4 & 3/4 \end{pmatrix} \begin{pmatrix} 1 \\ 1 \end{pmatrix}$$

$$= \begin{matrix} A_2 \\ A_3 \end{matrix} \begin{pmatrix} 1 \\ 1 \end{pmatrix}.$$

Assume now that we have an ergodic chain which satisfies the condition for lumpability for a partition **A**. The resulting chain will be ergodic. Let $\hat{A}$ be the limiting matrix for the lumped chain. Then we know that

$$\hat{A} = \lim_{n \to \infty} \frac{\hat{P} + \hat{P}^2 + \cdots + \hat{P}^n}{n}$$

$$\hat{A} = \lim_{n \to \infty} \frac{UPV + UP^2V + \cdots + UP^nV}{n}$$

$$\hat{A} = UAV.$$

In particular, this states that the components of $\hat{\alpha}$ are obtained from α by simply adding components in a given set. Similarly from the infinite series representation for the fundamental matrix $\hat{Z}$ we have

$$\hat{Z} = UZV.$$

There is in general no simple relation between M and $\hat{M}$. However, the mean time to go from a state in $\mathbf{A}_i$ to the set $\mathbf{A}_j$, in the original process, is the same for all states in $\mathbf{A}_i$. To see this we need only make the states of $\mathbf{A}_j$ absorbing. We know that the mean time to absorption is the same for all starting states chosen from a given set. If, in addition, $\mathbf{A}_j$ happens to consist of a single state, then $\hat{m}_{ij}$ may be found from M.

We can also compute the covariance matrix of the lumped process. As a matter of fact we know (see § 4.6.7) that the covariances are easily obtainable from C even if the original process is not lumpable with respect to the partition, that is, if the lumped process is not a Markov chain. In any case

$$\hat{C} = \sum_{\substack{s_k \text{ in } \mathbf{A}_i \\ s_l \text{ in } \mathbf{A}_j}} c_{kl}.$$

Let us carry out these computations for the Land of Oz example. For $\hat{A}$ we have

$$\hat{A} = \begin{pmatrix} 0 & 1 & 0 \\ 1/2 & 0 & 1/2 \end{pmatrix} \begin{pmatrix} 2/5 & 1/5 & 2/5 \\ 2/5 & 1/5 & 2/5 \\ 2/5 & 1/5 & 2/5 \end{pmatrix} \begin{pmatrix} 0 & 1 \\ 1 & 0 \\ 0 & 1 \end{pmatrix}$$

$$= \begin{pmatrix} 1/5 & 4/5 \\ 1/5 & 4/5 \end{pmatrix}$$

$$Z = \begin{pmatrix} 0 & 1 & 0 \\ 1/2 & 0 & 1/2 \end{pmatrix} \begin{pmatrix} 86/75 & 3/75 & -14/75 \\ 6/75 & 63/75 & 6/75 \\ -14/75 & 3/75 & 86/75 \end{pmatrix} \begin{pmatrix} 0 & 1 \\ 1 & 0 \\ 0 & 1 \end{pmatrix}$$

$$= \begin{pmatrix} 63/75 & 12/75 \\ 3/75 & 72/75 \end{pmatrix}$$

$$\hat{C} = \begin{pmatrix} 12/125 & -12/125 \\ -12/125 & 12/125 \end{pmatrix}$$

$$M = \begin{array}{c} \mathbf{R} \\ \mathbf{N} \\ \mathbf{S} \end{array} \overset{\begin{array}{ccc} \mathbf{R} & \mathbf{N} & \mathbf{S} \end{array}}{\begin{pmatrix} 5/2 & 4 & 10/3 \\ 8/3 & 5 & 8/3 \\ 10/3 & 4 & 5/2 \end{pmatrix}}.$$

From the fundamental matrix $\hat{Z}$ we find,

$$\hat{M} = \begin{array}{c} \mathbf{N} \\ \mathbf{B} \end{array} \overset{\begin{array}{cc} \mathbf{N} & \mathbf{B} \end{array}}{\begin{pmatrix} 5 & 1 \\ 4 & 5/4 \end{pmatrix}}.$$

Note that the mean time to reach **N** from either **R** or **S** is 4. Here **N** in the lumped process is a single element set. This common value is the mean time in the lumped chain to go from **B** to **N**. Similarly, the value 5 is obtainable from M. We observe that the mean time to go from **N** to **B** is considerably less than the mean time to go from **N** to either of the states in **B** in the original process.

§ **6.4 Weak lumpability.** In practice if one wanted to apply Markov chain ideas to a process for which the states have been combined, with respect to a partition $\mathbf{A} = \{\mathbf{A}_1, \mathbf{A}_2, \ldots, \mathbf{A}_n\}$, it is most natural to require that the resulting process be a Markov chain no matter what choice is made for the starting vector. However, there are some interesting theoretical considerations when we require only that at least one starting vector lead to a Markov chain. When this is the case we shall say that the process is *weakly lumpable with respect to the partition* **A**. We shall investigate the consequences of this weaker assumption in this section. We restrict the discussion to regular chains. The results of this section are based in part on results of C. K. Burke and M. Rosenblatt.†

For a given starting vector π, to determine whether or not the process is a Markov chain we must examine probabilities of the form

$$\mathbf{Pr}_\pi[\mathbf{f}_{n+1} \in \mathbf{A}_t | \mathbf{f}_n \in \mathbf{A}_s \wedge \cdots \wedge \mathbf{f}_1 \in \mathbf{A}_j \wedge \mathbf{f}_0 \in \mathbf{A}_i]. \qquad (1)$$

For a given π the process will be a Markov chain if these probabilities do not depend upon the outcomes before the n-th.

We must find conditions under which the knowledge of the outcomes before the last one does not affect the probability (1). Let us see how such knowledge could affect it. Given the information in (1), we know that after n steps the underlying chain is in a state in $\mathbf{A}_s$, but we do not know in which state it is. We can, however, assign probabilities for its being in each state of $\mathbf{A}_s$. We do this as follows: For any probability vector β, we denote by β^j the probability vector formed by making all components corresponding to states not in $\mathbf{A}_j$ equal to 0 and the remaining components proportional to those of β. We shall say that β^j is β *restricted to* $\mathbf{A}_j$. (If β has all components 0 in $\mathbf{A}_j$ we do not define β^j.) Consider now the information given in (1). The fact that $\mathbf{f}_0 \in \mathbf{A}_i$ may be interpreted as changing our initial vector to π^i. Learning then that $\mathbf{f}_1 \in \mathbf{A}_j$ may be interpreted as changing this vector to $(\pi^i P)^j$. We continue this process until we have taken into account all of the information given in (1). We are led to a certain assignment of probabilities for the states in $\mathbf{A}_s$. From these probabilities we can

† C. K. Burke and M. Rosenblatt, "A Markovian function of a Markov chain," *Annals of Mathematical Statistics*, **29**: 1112–1122, 1958.

easily compute the probability of a transition to $\mathbf{A}_t$ on the next step. But note that this probability may be quite different for different kinds of information. For example, our information may place high probability for being in a state from which it is certain that we move to $\mathbf{A}_t$. A different history of the process may place low probability on this state. These considerations give us a clue as to when we could expect that the past could be ignored. Two different cases are suggested. First would be the case where the information gained from the past would not do us any good. For example, assume that the probability for moving to the set $\mathbf{A}_t$ from a state in $\mathbf{A}_s$ is the same for all states in $\mathbf{A}_s$. Then clearly the probabilities for being in each state of $\mathbf{A}_s$ would not affect our predictions for the next outcome in the lumped process. This is the condition we found for lumpability in § 6.3.2. A second condition is suggested by the following: Assume that no matter what the past information is, we always end up with the same assignment of probabilities for being in each of the states in $\mathbf{A}_s$. Then again the past can have no influence on our predictions. We shall see that this case can also arise.

We have indicated above that the information given in (1) can be represented by a probability vector restricted to $\mathbf{A}_s$. This vector is obtained from the initial vector π by a sequence of transformations, each time taking into account one more bit of information. That is, we form the sequence

$$\left.\begin{aligned} \pi_1 &= \pi^i \\ \pi_2 &= (\pi_1 P)^j \\ \pi_3 &= (\pi_2 P)^k \\ \vdots &\quad\ \vdots \\ \pi_m &= (\pi_{m-1} P)^s \end{aligned}\right\} \tag{2}$$

We denote by $\mathbf{Y}_s$ the totality of vectors obtained by considering all finite sequences $\mathbf{A}_i, \mathbf{A}_j, \ldots, \mathbf{A}_s$, ending in $\mathbf{A}_s$.

6.4.1 THEOREM. *The lumped chain is a Markov chain for the initial vector π if and only if for every s and t the probability $\mathbf{Pr}_\beta[\mathbf{f}_1 \in \mathbf{A}_t]$ is the same for every β in $\mathbf{Y}_s$. This common value is the transition probability for moving from set $\mathbf{A}_s$ to set $\mathbf{A}_t$ in the lumped process.*

PROOF. The probability (1) can be represented in the form $\mathbf{Pr}_\beta[\mathbf{f}_1 \in \mathbf{A}_t]$ for a suitable β in $\mathbf{Y}_s$. To do this we use the first n outcomes for the construction (2). By hypothesis this probability depends only on s and t as required. Hence the lumped process is a Markov chain. Conversely, assume that the lumped chain is a Markov chain for initial vector π. Let β be any vector in $\mathbf{Y}_s$. Then β is obtained from a possible sequence, say of length $\mathbf{n}$, $\mathbf{A}_i, \mathbf{A}_j, \ldots, \mathbf{A}_s$. Let these be the given outcomes used to compute a probability of the form (1). This probability is $\mathbf{Pr}_\beta[\mathbf{f}_1 \in \mathbf{A}_t]$ and by the Markov property must not depend upon the outcomes before $\mathbf{A}_t$. Hence it has the same value for every β in $\mathbf{Y}_s$.

6.4.2 EXAMPLE. Consider a Markov chain with transition matrix

$$P = \begin{array}{c} \\ \mathbf{A}_1 \\ \\ \mathbf{A}_2 \end{array}\!\!\begin{array}{c} \begin{array}{ccc} \mathbf{A}_1 & & \mathbf{A}_2 \end{array} \\ \left(\begin{array}{c|cc} 1/4 & 1/4 & 1/2 \\ \hline 0 & 1/6 & 5/6 \\ 7/8 & 1/8 & 0 \end{array}\right) \end{array}.$$

Let $\mathbf{A} = (\{s_1\}, \{s_2, s_3\})$. Consider any vector of the form $(1-3a, a, 2a)$. Any such vector multiplied by P will again be of this form. Also any such vector restricted to $\mathbf{A}_1$ or $\mathbf{A}_2$ will be such a vector. Hence for any such starting vector the set $\mathbf{Y}_1$ will contain the single element $(1, 0, 0)$ and $\mathbf{Y}_2$ the single element $(0, 1/3, 2/3)$. Thus the condition of § 6.4.1 is satisfied trivially for any such starting vector. On the other hand assume that our starting vector is $\pi = (0, 0, 1)$. Let $\pi_1 = (\pi P)^2 = (0, 1, 0)$ and $\pi_2 = (\pi_1 P)^2 = (0, 1/6, 5/6)$. Then π_1 and π_2 are in $\mathbf{Y}_2$ and $\mathbf{Pr}_{\pi_1}[\mathbf{f}_1 \in \mathbf{A}_1] = 0$ while $\mathbf{Pr}_{\pi_2}[\mathbf{f}_1 \in \mathbf{A}_1] = 35/48$. Hence this choice of starting vector does not lead to a Markov chain.

We see that it is possible for certain starting vectors to lead to Markov chains while others do not. We shall now prove that if there is any starting vector which gives a Markov chain, then the fixed vector α does.

6.4.3 THEOREM. *Assume that a regular Markov chain is weakly lumpable with respect to* $\mathbf{A} = \{\mathbf{A}_1, \mathbf{A}_2, \ldots, \mathbf{A}_s\}$. *Then the starting vector* α *will give a Markov chain for the lumped process. The transition probabilities will be*

$$\hat{p}_{ij} = \mathbf{Pr}_{\alpha^i}[\mathbf{f}_1 \in \mathbf{A}_j].$$

Any other starting vector which yields a Markov chain for the lumped process will give the same transition probabilities.

PROOF. Since the chain is weakly lumpable there must be some starting vector π which leads to a Markov chain. Let its transition matrix be $\{\hat{p}_{ij}\}$. For this vector π

$$\mathbf{Pr}_\pi[\mathbf{f}_n \in \mathbf{A}_j | \mathbf{f}_{n-1} \in \mathbf{A}_i \wedge \mathbf{f}_{n-2} \in \mathbf{A}_k] = \hat{p}_{ij}$$

for all sets for which this probability is defined. But this may be written as

$$\mathbf{P}_{\pi P^{n-2}}[\mathbf{f}_2 \in \mathbf{A}_j | \mathbf{f}_1 \in \mathbf{A}_i \wedge \mathbf{f}_0 \in \mathbf{A}_k].$$

Letting n tend to infinity we have

$$\Pr_\alpha[\mathbf{f}_2 \in \mathbf{A}_j | \mathbf{f}_1 \in \mathbf{A}_i \wedge \mathbf{f}_0 \in \mathbf{A}_k] = \hat{p}_{ij}.$$

We have proved that the probability of the form (1), with α as starting vector, does not depend upon the past beyond the last outcome for the case $n = 1$. The general case is similar. Therefore, for α as a starting vector, the lumped process is a Markov chain. In the course of the proof we showed that $\hat{p}_{ij}$ for a starting vector π is the same as for α, hence it will be the same for any starting vector which yields a Markov chain.

By the previous theorem, if we are testing for weak lumpability we may assume that the process is started with the initial vector α. In this case the transition matrix $\hat{P}$ can be written in the form

$$\hat{P} = UPV$$

where V is as before but U is a matrix with i-th row α^i. When we have lumpability there is a great deal of freedom in the choice of U and in that case we chose a more convenient U. We do not have this freedom for weak lumpability.

We consider now conditions for which we can expect to have weak lumpability. If the chain is to be a Markov chain when lumped then we can compute $\hat{P}^2$ in two ways. Computing it directly from the underlying chain we have $\hat{P}^2 = UP^2V$. By squaring $\hat{P}$ we have $UPVUPV$. Hence it must be true that

$$UPVUPV = UPPV.$$

One sufficient condition for this is

$$VUPV = PV. \tag{3}$$

This is the condition for lumpability expressed in terms of our new U. It is necessary and sufficient for lumpability, and hence sufficient for weak lumpability. A second condition which would be sufficient for the above is

$$UPVU = UP. \tag{4}$$

This condition states the rows of UP are fixed vectors for VU. The matrix VU is now of the form

$$VU = \begin{pmatrix} W_1 & 0 & 0 \\ 0 & W_2 & 0 \\ 0 & 0 & W_3 \end{pmatrix},$$

where W_j is a transition matrix having all rows equal to α^j. To say that the i-th row of UP is a fixed vector for VU means that this vector, restricted to $\mathbf{A}_j$, is a fixed vector for W_j. But this means that the components of this vector must be proportional to α^j. Hence we have

$$(\alpha^i P)^j = \alpha^j. \tag{5}$$

This means that if we start with α, the set $\mathbf{Y}_i$, obtained by construction (2), consists, for each i, of a single element, namely α^i. Conversely, if each such set has only a single element, then (5) is satisfied and hence also (4). To say that $\mathbf{Y}_i$ has only one element for each i is to say that when the last outcome was $\mathbf{A}_i$ the knowledge of previous outcomes does not influence the assignment of the probabilities for being in each of the states of $\mathbf{A}_i$. Hence we have found that (4) is necessary and sufficient for the past beyond the last outcome to provide no new information, and is sufficient for weak lumpability.

Example 6.4.2 is a case where (4) is satisfied. Recall that we found that each $\mathbf{Y}_i$ had only one element.

We can summarize our findings as follows: We stated in the introduction that there are two obvious ways to make the information contained in the outcomes before the last one useless. One way is to require that even if we know the exact state of the original process our predictions would be unchanged. This is condition (3). The other is to require that we get no information at all from the past except the last step. This is condition (4). Each leads to weak lumpability. We have thus proved:

6.4.4. THEOREM. *Either condition* (3) *or condition* (4) *is sufficient for weak lumpability.*

There is an interesting connection between (3) and (4) in terms of the process and its associated reverse process (see § 5.3).

6.4.5 THEOREM. *A regular chain satisfies* (3) *if and only if the reverse chain satisfies* (4).

PROOF. Assume that a process satisfies (3). Then

$$VUPV = PV.$$

Let P_0 be the transition matrix for the reverse process, then $P = DP^T{}_0D^{-1}$. Hence

$$VUDP^T{}_0D^{-1}V = DP^T{}_0D^{-1}V$$

or, transposing,

$$V^TD^{-1}P_0DU^TV^T = V^TD^{-1}P_0D,$$

and

$$V^T D^{-1} P_0 D U^T V^T D^{-1} = V^T D^{-1} P_0.$$

We observe that $V^T D^{-1} = \hat{D}^{-1} U$. Furthermore, VUD is a symmetric matrix so that $VUD = DU^T V^T$ or $DU^T V^T D^{-1} = VU$. Using these two facts, our last equation becomes

$$\hat{D}^{-1} U P_0 V U = \hat{D}^{-1} U P_0.$$

Multiplying on the left by $\hat{D}$ gives condition (4) for P_0. The proof of the converse is similar.

6.4.6 THEOREM. *If a given process is weakly lumpable with respect to a partition* $\mathbf{A}$, *then so is the reverse process.*

PROOF. We must prove that all probabilities of the form

$$\mathbf{Pr}_\alpha[\mathbf{f}_1 \in \mathbf{A}_i | \mathbf{f}_2 \in \mathbf{A}_j \wedge \mathbf{f}_3 \in \mathbf{A}_h \wedge \cdots \wedge \mathbf{f}_n \in \mathbf{A}_t]$$

depend only on $\mathbf{A}_i$ and $\mathbf{A}_j$. We can write this probability in the form

$$\frac{\mathbf{Pr}_\alpha[\mathbf{f}_1 \in \mathbf{A}_i \wedge \mathbf{f}_2 \in \mathbf{A}_j \wedge \mathbf{f}_3 \in \mathbf{A}_h \wedge \cdots \wedge \mathbf{f}_n \in \mathbf{A}_t]}{\mathbf{Pr}_\alpha[\mathbf{f}_2 \in \mathbf{A}_j \wedge \mathbf{f}_3 \in \mathbf{A}_h \wedge \cdots \wedge \mathbf{f}_n \in \mathbf{A}_t]}$$

$$= \frac{\mathbf{Pr}_\alpha[\mathbf{f}_n \in \mathbf{A}_t \wedge \cdots \wedge \mathbf{f}_3 \in \mathbf{A}_h | \mathbf{f}_2 \in \mathbf{A}_j \wedge \mathbf{f}_1 \in \mathbf{A}_i]\ \mathbf{Pr}_\alpha[\mathbf{f}_1 \in \mathbf{A}_i \wedge \mathbf{f}_2 \in \mathbf{A}_j]}{\mathbf{Pr}_\alpha[\mathbf{f}_n \in \mathbf{A}_t \wedge \cdots \wedge \mathbf{f}_3 \in \mathbf{A}_h | \mathbf{f}_2 \in \mathbf{A}_j]\ \mathbf{Pr}_\alpha[\mathbf{f}_2 \in \mathbf{A}_j]}.$$

By hypothesis the forward process is a Markov chain, so that the first term in the numerator does not depend on $\mathbf{A}_i$. Hence this whole expression is simply

$$\frac{\mathbf{Pr}_\alpha[\mathbf{f}_1 \in \mathbf{A}_i \wedge \mathbf{f}_2 \in \mathbf{A}_j]}{\mathbf{Pr}_\alpha[\mathbf{f}_2 \in \mathbf{A}_j]}$$

which depends only on $\mathbf{A}_i$ and $\mathbf{A}_j$.

6.4.7 THEOREM. *A reversible regular Markov chain is reversible when lumped.*

PROOF. By reversibility,

$$P = DP^T D^{-1}$$

and

$$\hat{P} = UPV.$$

Hence

$$\hat{P} = UDP^T D^{-1} V.$$

We have seen that $V^TD^{-1} = \hat{D}^{-1}U$. Hence $UD = \hat{D}V^T$. Also $D^{-1}V = U^T\hat{D}^{-1}$. Thus we have

$$\hat{P} = \hat{D}V^TP^TU^T\hat{D}^{-1},$$

and hence

$$\hat{P} = \hat{D}\hat{P}^T\hat{D}^{-1}.$$

This means that the lumped process is reversible.

6.4.8 Theorem. *For a reversible regular Markov chain, weak lumpability implies lumpability.*

PROOF. Let P be the transition matrix for a regular reversible chain. Then, if the chain is weakly lumpable,

$$UPPV = UPVUPV$$

or

$$UP(I - VU)PV = 0.$$

Since $U = \hat{D}V^TD^{-1}$, we have

$$\hat{D}V^TD^{-1}P(I - VU)PV = 0,$$

or, multiplying through by $\hat{D}^{-1}$ and using the fact that for a reversible chain $D^{-1}P = P^TD^{-1}$, we have

$$V^TP^TD^{-1}(I - VU)PV = 0.$$

Let $W = D^{-1} - D^{-1}VU$. Then $W = D^{-1} - U^T\hat{D}^{-1}U$. We shall show that W is semi-definite. That is, for any vector β, $\beta^TW\beta$ is non-negative. It is sufficient to prove that

$$\sum_{k \text{ in } A_i} a_kb^2_k \geqslant \hat{d}_i\left(\sum_{k \text{ in } A_i} a_kb_k\right)^2$$

where $\hat{d}_i$ is the i-th diagonal entry of $\hat{D}$, or equivalently

$$\sum_{k \text{ in } A_i} a_k\hat{d}_ib^2_k \geqslant \sum_{k \text{ in } A_i} (a_k\hat{d}_ib_k)^2.$$

But since the coefficients $a_k\hat{d}_i$ are non-negative and have sum 1, this is a standard inequality of probability theory. It can be proved by considering a function **f** which takes on the value b_k with probability $a_k\hat{d}_i$. Then the inequality expresses that $\mathbf{M}[\mathbf{f}^2] \geqslant (\mathbf{M}[\mathbf{f}])^2$; and, by § 1.8.5, this simply asserts that the variance of **f** is non-negative.

Since W_i is semi-definite, $W = X^TX$ for some matrix X. Thus

$$V^TP^TX^TXPV = 0$$

or

$$(XPV)^T(XPV) = 0.$$

This can be true only if

$$XPV = 0.$$

Hence

$$X^T XPV = 0,$$

or

$$D^{-1}(I - VU)PV = 0,$$

or

$$(I - VU)PV = 0.$$

Hence

$$PV = VUPV.$$

Note that while we have given necessary and sufficient conditions for lumpability with respect to a partition **A**, we have not given necessary and sufficient conditions for weak lumpability. We have given two different sufficient conditions (3) and (4). It might be hoped that for weak lumpability one of the two conditions would have to be satisfied. It is, however, easy to get an example where neither is satisfied as follows: If we take a Markov chain and find a method of combining states to give a Markov chain, we can then ask whether the new chain can be combined. If so, the result can be considered a combining of states in the original chain. To get our counterexample, we take a chain for which we can combine states by condition (3) and then combine states in the new chain by condition (4); the result considered as a lumping of the original chain will obviously be a Markov chain, but it will satisfy neither (4) nor (3). Consider a Markov chain with transition matrix

$$P = \begin{array}{c} \mathbf{A}_1 \\ \mathbf{A}_2 \\ \\ \mathbf{A}_3 \end{array} \left(\begin{array}{c|cc|c} 1/4 & 1/16 & 3/16 & 1/2 \\ \hline 0 & 1/12 & 1/12 & 5/6 \\ 0 & 1/12 & 1/12 & 5/6 \\ \hline 7/8 & 1/32 & 3/32 & 0 \end{array} \right)$$

For the partition $\mathbf{A} = (\{s_1\}, \{s_2, s_3\}, \{s_4\})$ the strong condition (3) is satisfied. Hence we obtain a lumped chain with transition matrix

$$P = \begin{array}{c} \\ \mathbf{A}_1 \\ \mathbf{A}_2 \\ \mathbf{A}_3 \end{array} \begin{array}{c} \begin{array}{ccc} \mathbf{A}_1 & \mathbf{A}_2 & \mathbf{A}_3 \end{array} \\ \begin{pmatrix} 1/4 & 1/4 & 1/2 \\ 0 & 1/6 & 5/6 \\ 7/8 & 1/8 & 0 \end{pmatrix} \end{array}.$$

But this is Example 6.4.2, which satisfies (4). Hence we can lump it by $(\{\mathbf{A}_1\}, \{\mathbf{A}_2, \mathbf{A}_3\})$. The result is a lumping of the original chain by

$\mathbf{A} = (\{s_1\}, \{s_2, s_3, s_4\})$. It is easily checked that neither (3) nor (4) is satisfied in the original process for this partition.

We conclude with some remarks about a lumped process when the condition for weak lumpability is not satisfied. We assume that P is regular. Then if the process is started in equilibrium

$$\hat{p}_{ij} = \mathbf{Pr}_\alpha[\mathbf{f}_{n+1} \in \mathbf{A}_j | \mathbf{f}_n \in \mathbf{A}_i]$$

is the same for every n. Hence the matrix $\hat{P} = \{\hat{p}_{ij}\}$ may still be interpreted as a one-step transition matrix. Also

$$\hat{a}_i = \mathbf{Pr}_\alpha[\mathbf{f}_n \in \mathbf{A}_i]$$

is the same for all n. The vector $\hat{a} = \{\hat{a}_i\}$ will be the unique fixed vector for $\hat{P}$. Its components may be obtained from α by simply adding the components corresponding to each set. Similarly we may define two-step transition probabilities by

$$\hat{p}^{(2)}{}_{ij} = \mathbf{Pr}_\alpha[\mathbf{f}_{n+2} \in \mathbf{A}_j | \mathbf{f}_n \in \mathbf{A}_i].$$

The two-step transition matrix will then be $\hat{P}^{(2)} = \{\hat{p}^{(2)}{}_{ij}\}$. It will no longer be true that $\hat{P}^2 = \hat{P}^{(2)}$.

We can also define the mean first passage matrix $\hat{M}$ for the lumped process. It cannot be obtained by our Markov chain formulas. To obtain $\hat{M}$ it is necessary first to find m_{i,A_j}, the mean time to go from state i to set $\mathbf{A}_j$ in the original process. We can do this by making all of the elements of $\mathbf{A}_j$ absorbing and find the mean time to absorption. (A slight modification is necessary if i is in $\mathbf{A}_j$.) From these we obtain the mean time to go from $\mathbf{A}_i$ to $\mathbf{A}_j$, by

$$\hat{m}_{ij} = \sum_{k \text{ in } \mathbf{A}_i} a^*{}_k m_{k,A_j}$$

where $a^*{}_k$ is the k-th component of α^i.

§ 6.5 Expanding a Markov chain. In the last two sections we showed that under certain conditions a Markov chain would, by lumping states together, be reduced to a smaller chain which gave interesting information about the original chain. By this process we obtained a more manageable chain at the sacrifice of obtaining less precise information. In this section we shall show that it is possible to go in the other direction. That is, to obtain from a Markov chain a larger chain which gives more detailed information about the process being considered. We shall base the presentation on results obtained by S. Hudson in his senior thesis at Dartmouth College.

Consider now a Markov chain with states $s_1, s_2, \ldots, s_r$. We form a new Markov chain, called the *expanded process*, as follows. A state is a pair of states (s_i, s_j) in the original chain, for which $p_{ij} > 0$. We denote these states by $s_{(ij)}$. Assume now that in the original chain the transition from s_i to s_j and from s_j to s_k occurs on two successive steps. We shall interpret this as a single step in the expanded process from the state $s_{(ij)}$ to the state $s_{(jk)}$. With this convention, transition from state $s_{(ij)}$ to state $s_{(kl)}$ in the expanded process is possible only if $j = k$. Transition probabilities are given by

$$p_{(ij)(jl)} = p_{jl},$$
$$p_{(ij)(kl)} = 0 \quad \text{for} \quad j \neq k.$$

Or

$$p_{(ij)(kl)} = p_{jl} \cdot d_{jk}.$$

6.5.1 EXAMPLE. Consider the Land of Oz example. The states for the expanded process are **RR**, **RN**, **RS**, **NR**, **NS**, **SR**, **SN**, **SS**. Note that **NN** is not a state, since $p_{NN} = 0$ in the original process. The transition matrix for the expanded process is

$$\begin{array}{c} \\ \mathbf{RR} \\ \mathbf{RN} \\ \mathbf{RS} \\ \mathbf{NR} \\ \mathbf{NS} \\ \mathbf{SR} \\ \mathbf{SN} \\ \mathbf{SS} \end{array}
\begin{array}{c} \begin{array}{cccccccc} \mathbf{RR} & \mathbf{RN} & \mathbf{RS} & \mathbf{NR} & \mathbf{NS} & \mathbf{SR} & \mathbf{SN} & \mathbf{SS} \end{array} \\
\begin{pmatrix}
1/2 & 1/4 & 1/4 & 0 & 0 & 0 & 0 & 0 \\
0 & 0 & 0 & 1/2 & 1/2 & 0 & 0 & 0 \\
0 & 0 & 0 & 0 & 0 & 1/4 & 1/4 & 1/2 \\
1/2 & 1/4 & 1/4 & 0 & 0 & 0 & 0 & 0 \\
0 & 0 & 0 & 0 & 0 & 1/4 & 1/4 & 1/2 \\
1/2 & 1/4 & 1/4 & 0 & 0 & 0 & 0 & 0 \\
0 & 0 & 0 & 1/2 & 1/2 & 0 & 0 & 0 \\
0 & 0 & 0 & 0 & 0 & 1/4 & 1/4 & 1/2
\end{pmatrix} \end{array}.$$

Let us first see how the classification of states for the expanded process compares with the original chain. We note that $p^{(n)}{}_{(ij)(kl)} = p^{(n-1)}{}_{jk} p_{kl} > 0$ if and only if $p^{(n-1)}{}_{jk} > 0$. Hence if the original chain is ergodic, so will the expanded process be, and if the original chain is of period d, then the expanded chain will also be of period d. A state $s_{(ij)}$ in the expanded process is absorbing only if $i = j$ and only if state s_j is absorbing in the original chain.

Assume that the original chain is an absorbing chain. Let $s_{(ij)}$ be a non-absorbing state in the expanded process. Since the original chain was absorbing, there must be an absorbing state s_k such that it is possible to go from s_j to s_k. Thus it is possible to go from $s_{(ij)}$ to $s_{(kk)}$ in the expanded process. Thus the expanded process is also absorbing.

It is interesting to observe that from the expanded process we can go back to the original process by lumping states. For this we form the partition $\mathbf{A} = \{\mathbf{A}_1, \mathbf{A}_2, \ldots, \mathbf{A}_r\}$ of the states in the extended chain, with $\mathbf{A}_i$ the set of all states of the form $\mathbf{s}_{(ki)}$. Then the condition for lumping is that $p_{(ki)A_j}$ should not depend on k. But this is true by the Markov property for the original chain. The lumped process is then the same as the original chain. In our example, the partition is

$$\mathbf{A} = \{(\mathbf{RR}, \mathbf{SR}, \mathbf{NR}), (\mathbf{SN}, \mathbf{RN}), (\mathbf{RS}, \mathbf{NS}, \mathbf{SS})\}.$$

We next compare the basic quantities for our expanded process with the corresponding quantities for the original chain. We shall treat only the regular case. The other cases may be treated similarly.

6.5.2 THEOREM. *Let $\alpha = \{a_i\}$ be the fixed vector for a regular chain with transition matrix P. Let $\hat{\alpha} = \{a_{(ij)}\}$ be the fixed vector for the expanded chain. Then*

$$a_{(ij)} = a_i p_{ij}.$$

PROOF. It is obvious that $a_i p_{ij}$ is positive. Also,

$$\sum_{(ij)} a_{(ij)} = \sum_{i,j} a_i p_{ij} = \sum_j a_j = 1.$$

Hence we need only prove that $\hat{\alpha} = \{a_i p_{ij}\}$ is a fixed vector for the transition matrix for the expanded process. That is,

$$\sum_{(ij)} a_{(ij)} p_{(ij)(kl)} = a_{(kl)}.$$

But

$$\begin{aligned}
\sum_{(ij)} a_{(ij)} p_{(ij)(kl)} &= \sum_{i,j} a_i p_{ij} p_{jl} d_{jk} \\
&= \sum_j a_j p_{jl} d_{jk} \\
&= a_k p_{kl} \\
&= a_{(kl)}.
\end{aligned}$$

In our example, this gives for the fixed vector

$$\hat{\alpha} = (.2, .1, .1, .1, .1, .1, .1, .2).$$

Note that the result we have proved is intuitively obvious, since $a_{(ij)}$ represents the probability that after a large number of steps the process will be in state $\mathbf{s}_i$ and then move to state $\mathbf{s}_j$. The probability that this will occur is clearly $a_i p_{ij}$.

6.5.3 THEOREM. *The fundamental matrix for the expanded chain is*

$$\hat{Z} = \{z_{(ij)(kl)}\} = \{d_{(ij)(kl)} + (z_{jk} - a_k) p_{kl}\}.$$

PROOF.

$$
\begin{aligned}
z_{(ij)(kl)} &= d_{(ij)(kl)} + \sum_{n=1}^{\infty} (p^{(n)}{}_{(ij)(kl)} - a_{(kl)}) \\
&= d_{(ij)(kl)} + \sum_{n=1}^{\infty} (p^{(n-1)}{}_{jk} p_{kl} - a_k p_{kl}) \\
&= d_{(ij)(kl)} + p_{kl} \sum_{n=0}^{\infty} (p^{(n)}{}_{jk} - a_k) \\
&= d_{(ij)(kl)} + p_{kl}(z_{jk} - a_k).
\end{aligned}
$$

For our example

$$
\hat{Z} = \begin{array}{c} \\ \text{RR} \\ \text{RN} \\ \text{RS} \\ \text{NR} \\ \text{NS} \\ \text{SR} \\ \text{SN} \\ \text{SS} \end{array}
\begin{array}{c}
\begin{array}{cccccccc} \text{RR} & \text{RN} & \text{RS} & \text{NR} & \text{NS} & \text{SR} & \text{SN} & \text{SS} \end{array} \\
\begin{pmatrix}
1.373 & .187 & .187 & -.080 & -.080 & -.147 & -.147 & -.293 \\
-.160 & .920 & -.080 & .320 & .320 & -.080 & -.080 & -.160 \\
-.293 & -.147 & .853 & -.080 & -.080 & .187 & .187 & .373 \\
.373 & .187 & .187 & .920 & -.080 & -.147 & -.147 & -.293 \\
-.293 & -.147 & -.147 & -.080 & .920 & .187 & .187 & .373 \\
.373 & .187 & .187 & -.080 & -.080 & .853 & -.147 & -.293 \\
-.160 & -.080 & -.080 & .320 & .320 & -.080 & .920 & -.160 \\
-.293 & -.147 & -.147 & -.080 & -.080 & .187 & .187 & 1.373
\end{pmatrix}
\end{array}.
$$

We next consider the mean first passage times for the expanded process.

6.5.4 THEOREM.

$$
m_{(ij)(kl)} = \frac{1}{a_k p_{kl}} - \frac{(z_{jk} - z_{lk})}{a_k}.
$$

PROOF. From the matrix expression for M in terms of the fundamental matrix we have

$$
\begin{aligned}
m_{(ij)(kl)} &= (d_{(ij)(kl)} - z_{(ij)(kl)} + z_{(kl)(kl)}) \frac{1}{a_{(kl)}} \\
&= [d_{(ij)(kl)} - p_{kl}(z_{jk} - a_k) \\
&\qquad - d_{(ij)(kl)} + p_{kl}(z_{lk} - a_k) + 1] \frac{1}{a_k p_{kl}} \\
&= [1 - p_{kl}(z_{jk} - z_{lk})] \frac{1}{a_k p_{kl}} \\
&= \frac{1}{a_k p_{kl}} - \frac{(z_{jk} - z_{lk})}{a_k}.
\end{aligned}
$$

Again, as was to be expected, $m_{(ij)(kl)}$ does not depend on i.

For our example, we obtain

$$\hat{M} = \begin{array}{c} \\ \mathbf{RR} \\ \mathbf{RN} \\ \mathbf{RS} \\ \mathbf{NR} \\ \mathbf{NS} \\ \mathbf{SR} \\ \mathbf{SN} \\ \mathbf{SS} \end{array}\begin{array}{c} \begin{array}{cccccccc} \mathbf{RR} & \mathbf{RN} & \mathbf{RS} & \mathbf{NR} & \mathbf{NS} & \mathbf{SR} & \mathbf{SN} & \mathbf{SS} \end{array} \\ \begin{pmatrix} 5 & 7^1/_3 & 6^2/_3 & 10 & 10 & 10 & 10^2/_3 & 8^1/_3 \\ 7^2/_3 & 10 & 9^1/_3 & 6 & 6 & 9^1/_3 & 10 & 7^2/_3 \\ 8^1/_3 & 10^2/_3 & 10 & 10 & 10 & 6^2/_3 & 7^1/_3 & 5 \\ 5 & 7^1/_3 & 6^2/_3 & 10 & 10 & 10 & 10^2/_3 & 8^1/_3 \\ 8^1/_3 & 10^2/_3 & 10 & 10 & 10 & 6^2/_3 & 7^1/_3 & 5 \\ 5 & 7^1/_3 & 6^2/_3 & 10 & 10 & 10 & 10^2/_3 & 8^1/_3 \\ 7^2/_3 & 10 & 9^1/_3 & 6 & 6 & 9^1/_3 & 10 & 7^2/_3 \\ 8^1/_3 & 10^2/_3 & 10 & 10 & 10 & 6^2/_3 & 7^1/_3 & 5 \end{pmatrix} \end{array}.$$

In comparison, the mean first passage matrix for the original chain is

$$M = \begin{array}{c} \\ \mathbf{R} \\ \mathbf{N} \\ \mathbf{S} \end{array}\begin{array}{c} \begin{array}{ccc} \mathbf{R} & \mathbf{N} & \mathbf{S} \end{array} \\ \begin{pmatrix} 2.5 & 4 & 3.3 \\ 2.7 & 5 & 2.7 \\ 3.3 & 4 & 2.5 \end{pmatrix} \end{array}.$$

Consider next the reverse transition matrix for the expanded process. The transition probabilities are

$$\hat{p}_{(ij)(kl)} = \frac{a_{(kl)}p_{(kl)(ij)}}{a_{(ij)}}.$$

Hence

$$\hat{p}_{(ij)(kl)} = 0 \quad \text{if } i \neq l$$

and

$$\begin{aligned} \hat{p}_{(ij)(ki)} &= \frac{a_{(ki)}p_{(ki)(ij)}}{a_{(ij)}} \\ &= \frac{a_k p_{ki} p_{ij}}{a_i p_{ij}} \\ &= \frac{a_k p_{ki}}{a_i} \\ &= \hat{p}_{ik}. \end{aligned}$$

Hence the reverse process for the expanded process is simply the reverse process for the original chain expanded.

One application of the expanded process is the following: It often happens that the transition matrix P for a chain is not known and must be estimated from data. If a large number n of outcomes for the process are known, then an obvious estimate is

$$p_{ij} = \frac{\mathbf{y}^{(n)}{}_{ij}}{\mathbf{y}^{(n)}{}_i}$$

where $\mathbf{y}^{(n)}{}_{ij}$ is the number of transitions from state $\mathbf{s}_i$ to state $\mathbf{s}_j$, and $\mathbf{y}^{(n)}{}_i$ is the number of times the process is in state $\mathbf{s}_i$. To study the properties of this estimate it is necessary to study the properties of $\mathbf{y}^{(n)}{}_{ij}$ and $\mathbf{y}^{(n)}{}_i$ for a Markov chain. In particular, the limiting variances and covariances for these quantities are important. We know that we can obtain the limiting covariances for $\mathbf{y}^{(n)}{}_i$, $\mathbf{y}^{(n)}{}_j$ from the fundamental matrix for the basic chain. But how do we obtain the limiting covariances for $\mathbf{y}^{(n)}{}_{(ij)}$, $\mathbf{y}^{(n)}{}_{(kl)}$? We simply observe that these are the limiting covariances for the number of times in a pair of states for the expanded chain. Hence we can express these covariances in terms of $\hat{Z}$ and $\hat{a}$ for the expanded process. We can then use Theorems 6.5.2, 6.5.3 to express the limiting covariances in terms of quantities relating to the original chain. Carrying out this computation gives:

6.5.5 THEOREM. *The limiting covariances for the expanded chains are given by*

$$c_{(ij)(kl)} = a_i p_{ij} p_{kl} z_{jk} + a_k p_{kl} p_{ij} z_{li} + a_k p_{kl} d_{(ij)(kl)} - 3 a_i p_{ij} a_k p_{kl}.$$

For our example these covariances are

$$\hat{C} = \begin{pmatrix} .309 & .001 & -.012 & .001 & -.065 & -.012 & -.065 & -.157 \\ .001 & .074 & -.033 & -.041 & .007 & .001 & -.026 & -.065 \\ -.012 & -.033 & .061 & .001 & -.033 & .027 & .001 & -.012 \\ .001 & .041 & .001 & .074 & -.026 & -.033 & .007 & -.065 \\ -.065 & .007 & -.033 & -.026 & .074 & .001 & .041 & .001 \\ -.012 & .001 & .027 & -.033 & .001 & .061 & -.033 & -.012 \\ -.065 & -.026 & .001 & .007 & .041 & -.033 & .074 & .001 \\ -.157 & -.065 & -.012 & -.065 & .001 & -.012 & .001 & .309 \end{pmatrix}.$$

Exercises for Chapter VI

For § 6.1

1. Consider Example 2 with $p = {}^1/_2$. Assume that the process is observed only when it is in the set $\{\mathbf{s}_2, \mathbf{s}_3, \mathbf{s}_4\}$. Find the resulting transition matrix. Find M for the new process. What do the entries of M mean in terms of the original chain?

2. The following table gives the probability of a team ending up in a certain position next year, given what its position is this year. For each position in the second division, calculate the mean number of years to reach the first division.

		1	2	3	4	5	6	7	8
FIRST DIVISION	1ST	.3	.3	.3	.1	0	0	0	0
	2ND	.1	.2	.2	.2	.2	.1	0	0
	3RD	.1	.1	.2	.2	.1	.1	.1	.1
	4TH	0	.1	.1	.2	.2	.2	.1	.1
SECOND DIVISION	5TH	0	.05	.05	.2	.3	.2	.1	.1
	6TH	0	0	.1	.1	.2	.3	.2	.1
	7TH	0	0	0	0	.1	.3	.3	.3
	8TH	0	0	0	0	.1	.2	.4	.3

3. Consider a chain with a single ergodic set, which has transient states. Let $\mathbf{s}_i$ be a transient state and $\mathbf{s}_j$ an ergodic state. Let $\mathbf{f}_j$ be the time required to reach $\mathbf{s}_j$, $\mathbf{g}$ the first ergodic state reached and $\mathbf{t}$ the time required to reach the ergodic set. Then

$$\mathbf{M}_i[\mathbf{f}_j] = \sum_{\mathbf{s}_k \text{ ergodic}} \mathbf{Pr}_i[\mathbf{g} = \mathbf{s}_k][\mathbf{M}_k[\mathbf{f}_j] + \mathbf{M}_i[\mathbf{t}|\mathbf{g} = s_k]].$$

Use this result to find, for Example 4 with $p = {}^2/_3$, the mean time to reach state $\mathbf{s}_1$ for the first time starting in state $\mathbf{s}_3$.

4. It is raining in the Land of Oz. Find the mean number of days until each kind of weather has occurred at least once.

5. Prove that when a process is observed only when in a subset of an ergodic chain, the resulting process is also an ergodic chain.

For § 6.2

6. Find the mean number of rainy days between nice days in the Land of Oz.

7. For the Markov chain in Exercise 2 of Chapter II, assume that when the duel ends a new duel is started. Find the fixed vector for the resulting chain. Use this to determine the absorption probabilities and the mean number of times in each state for the original chain.

8. Consider the chain with transition matrix

$$P = \begin{matrix} & \mathbf{s}_1 & \mathbf{s}_2 & \mathbf{s}_3 \\ \mathbf{s}_1 & 0 & 1 & 0 \\ \mathbf{s}_2 & {}^1/_4 & 0 & {}^3/_4 \\ \mathbf{s}_3 & 0 & 1 & 0 \end{matrix}.$$

Find the fundamental matrix Z by making state $\mathbf{s}_1$ into an absorbing state, calculating N, and using the result of Theorem 6.2.5. Find $m_{12} + m_{21}$ by using Corollary 6.2.7.

9. From 6.2.5 deduce the identity

$$I - A = (I - P)N^*(I - A).$$

For § 6.3

10. For Example 2 with $p = {}^1/_2$, which of the following partitions produces a Markov chain when lumped?

(a) $A = (\{s_1, s_3, s_5\}, \{s_2, s_4\})$.
(b) $B = (\{s_1, s_5\}, \{s_2, s_4\}, \{s_3\})$.

Which produce Markov chains if $p \neq {}^1/_2$?

11. Show that Example 3(*a*) is lumpable with respect to the partition $A = (\{s_1, s_5\}, \{s_2, s_4\}, \{s_3\})$. Find the fundamental matrix for the lumped chain from the fundamental matrix for the original chain (see § 4.7).

12. Find a three-cell partition which makes Example 6 lumpable.

13. Let P be the transition matrix of an independent trials chain and a be any number with $0 < a < 1$. Show that $P' = aP + (1-a)I$ is lumpable with respect to any partition.

14. Show that Example 12 is lumpable with respect to the partition $A = (\{s_1s_1, s_2s_1\}, \{s_1s_2, s_2s_2\})$. How is the resulting transition matrix related to the two-state chain which determined the four-state chain?

15. Prove that for a lumpable ergodic chain, $\hat{\alpha} = \alpha V$.

16. Give an example of a Markov chain which is not itself an independent trials chain, but which can be lumped to an independent trials chain. Check your answer by computing Z and UZV.

For § 6.4

17. Show that the Markov chain with transition matrix

$$\begin{matrix} s_1 \\ s_2 \\ s_3 \end{matrix} \begin{pmatrix} 1/2 & 1/4 & 1/4 \\ 1/2 & 1/2 & 0 \\ 0 & 1/4 & 3/4 \end{pmatrix}$$

is weakly lumpable, but not lumpable, with respect to $A = (\{s_1\}, \{s_2, s_3\})$. Find the transition matrix for the lumped process. Show that the reverse process is lumpable.

18. Show that the Markov chain with transition matrix

$$P = \begin{matrix} s_1 \\ s_2 \\ s_3 \\ s_4 \\ s_5 \end{matrix} \begin{pmatrix} 0 & 3/4 & 1/4 & 0 & 0 \\ 3/4 & 0 & 1/4 & 0 & 0 \\ 0 & 1/8 & 5/8 & 1/8 & 1/8 \\ 1/16 & 1/16 & 1/8 & 1/4 & 1/2 \\ 1/8 & 0 & 1/8 & 3/8 & 3/8 \end{pmatrix}$$

is weakly lumpable with respect to $(\{s_1, s_2, s_3\}, \{s_4, s_5\})$.

19. For the Land of Oz example, let $\mathbf{A}_1 = \{\mathbf{R}, \mathbf{N}\}$ and $\mathbf{A}_2 = \{\mathbf{S}\}$. Compute $\mathbf{Pr}_\alpha[\mathbf{f}_2 \in \mathbf{A}_1 | \mathbf{f}_1 \in \mathbf{A}_1]$ and $\mathbf{Pr}_\alpha[\mathbf{f}_2 \in \mathbf{A}_1 | \mathbf{f}_1 \in \mathbf{A}_1 \wedge \mathbf{f}_0 \in \mathbf{A}_1]$. Use the result to show that the chain is not weakly lumpable with respect to $\mathbf{A} = (\mathbf{A}_1, \mathbf{A}_2)$.

20. Prove that if P is lumpable with respect to a given partition, and if P has column sums 1, then P^T is weakly lumpable with respect to the same partition.

For § 6.5

21. A coin is tossed a sequence of times. Represent this as a two-state Markov chain. Form the expanded process. Find M and M_2 for the expanded chain. Interpret the diagonal entries in terms of the original chain.

22. Let P be the transition matrix of an independent trials chain. Find formulas for Z and M of the expanded chain. Check the latter against Exercise 21.

23. Let P be the transition matrix of an independent trials chain. Find a formula for the limiting covariances of the expanded chain. [HINT: Write the formula in the form $a_k a_l(\cdots)$.] Use this formula to compute the limiting covariances in Exercise 21.

CHAPTER VII

APPLICATIONS OF MARKOV CHAINS

§ 7.1 **Random walks.** We will consider four simple, related random walks. The first three are walks on a line, with states $0, 1, \ldots, n$:

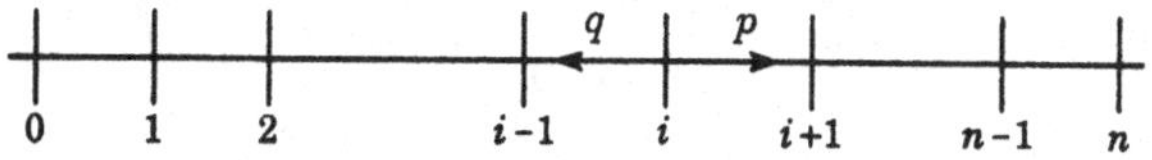

FIGURE 7-1

In each of the first three types of random walks we have probability p of moving to the right (from i to $i+1$) and probability q of moving to the left (from i to $i-1$), for states $i = 1, 2, \ldots, n-1$. The three types differ in their behavior at the "boundaries," 0 and n.

AARW: A random walk having both 0 and n as absorbing states.

APRW: A random walk having 0 as an absorbing state, while n is "partially reflecting." That is, at n it moves back to $n-1$ with probability q and stays at n with probability p.

PPRW: A random walk partially reflecting at both boundaries. That is it is like APRW at n, and at 0 it moves to 1 with probability p and stays at 0 with probability q.

The fourth random walk will move on a circle, with states numbered 1, 2, $\ldots, n$, as in Figure 7-2.

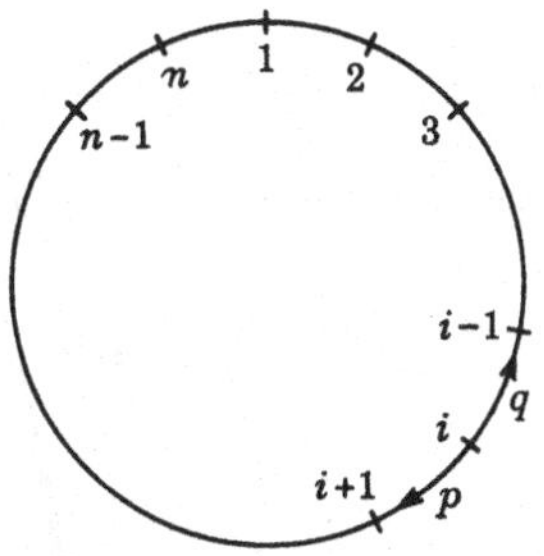

FIGURE 7-2

CRW: The process moves on the circle, taking one step clockwise with probability p, and counterclockwise with probability q.

We will illustrate each of these random walks for $n=5$. Since the behavior is often quite different for $p={}^1/_2$ than for any other p-value, we will carry out the illustration for both $p={}^1/_2$ and $p={}^2/_3$. Then we will solve each random walk. *It will be convenient to let r stand for p/q.*

Let us first consider AARW. This is an absorbing chain with two absorbing states 0 and n.

AARW for $n=5$, $p={}^1/_2$

$$P = \begin{array}{c} \\ 0 \\ 5 \\ 1 \\ 2 \\ 3 \\ 4 \end{array} \begin{array}{c} \begin{array}{cccccc} 0 & 5 & 1 & 2 & 3 & 4 \end{array} \\ \left(\begin{array}{cc|cccc} 1 & 0 & 0 & 0 & 0 & 0 \\ 0 & 1 & 0 & 0 & 0 & 0 \\ \hline 1/2 & 0 & 0 & 1/2 & 0 & 0 \\ 0 & 0 & 1/2 & 0 & 1/2 & 0 \\ 0 & 0 & 0 & 1/2 & 0 & 1/2 \\ 0 & 1/2 & 0 & 0 & 1/2 & 0 \end{array}\right) \end{array}$$

$$N = \begin{array}{c} 1 \\ 2 \\ 3 \\ 4 \end{array} \begin{pmatrix} 1.6 & 1.2 & .8 & .4 \\ 1.2 & 2.4 & 1.6 & .8 \\ .8 & 1.6 & 2.4 & 1.2 \\ .4 & .8 & 1.2 & 1.6 \end{pmatrix} \qquad \tau = \begin{pmatrix} 4 \\ 6 \\ 6 \\ 4 \end{pmatrix}$$

$$B = \begin{array}{c} \\ 1 \\ 2 \\ 3 \\ 4 \end{array} \begin{array}{c} \begin{array}{cc} 0 & 5 \end{array} \\ \begin{pmatrix} .8 & .2 \\ .6 & .4 \\ .4 & .6 \\ .2 & .8 \end{pmatrix} \end{array}$$

AARW for $n=5$, $p={}^2/_3$

$$P = \begin{array}{c} \\ 0 \\ 5 \\ 1 \\ 2 \\ 3 \\ 4 \end{array} \begin{array}{c} \begin{array}{cccccc} 0 & 5 & 1 & 2 & 3 & 4 \end{array} \\ \left(\begin{array}{cc|cccc} 1 & 0 & 0 & 0 & 0 & 0 \\ 0 & 1 & 0 & 0 & 0 & 0 \\ \hline 1/3 & 0 & 0 & 2/3 & 0 & 0 \\ 0 & 0 & 1/3 & 0 & 2/3 & 0 \\ 0 & 0 & 0 & 1/3 & 0 & 2/3 \\ 0 & 2/3 & 0 & 0 & 1/3 & 0 \end{array}\right) \end{array}$$

$$N = {}^1/_{31}\begin{pmatrix} 45 & 42 & 36 & 24 \\ 21 & 63 & 54 & 36 \\ 9 & 27 & 63 & 42 \\ 3 & 9 & 21 & 45 \end{pmatrix}$$

$$\tau = {}^1/_{31}\begin{pmatrix} 147 \\ 174 \\ 141 \\ 78 \end{pmatrix} \qquad B = {}^1/_{31}\begin{matrix} & \begin{matrix} 0 & 5 \end{matrix} & \\ & \begin{pmatrix} 15 & 16 \\ 7 & 24 \\ 3 & 28 \\ 1 & 30 \end{pmatrix} & \begin{matrix} 1 \\ 2 \\ 3 \\ 4 \end{matrix} \end{matrix}$$

The matrix $I-Q$ has the form

$$I-Q = \begin{pmatrix} 1 & -p & 0 & 0 \\ -q & 1 & -p & 0 \\ 0 & -q & 1 & -p \\ 0 & 0 & -q & 1 \end{pmatrix}$$

when $n=5$. In general it has entries s_{ij} $(i, j=1, 2, \ldots, n-1)$ which are 0 except that $s_{jj}=1$, $s_{j-1,j}=-p$, $s_{j+1,j}=-q$. We note from the numerical examples that the entries of N decrease on both sides of the diagonal. N will have the form

$$n_{ij} = \frac{1}{(p-q)(r^n-1)}\cdot\begin{cases}(r^j-1)(r^{n-i}-1) & \text{if } j \leqslant i \\ (r^i-1)(r^{n-i}-r^{j-i}) & \text{if } j \geqslant i\end{cases} \tag{1}$$

except where $p={}^1/_2$, here the solution simplifies to

$$n_{ij} = \frac{2}{n}\cdot\begin{cases} j(n-i) & \text{if } j \leqslant i \\ i(n-j) & \text{if } j \geqslant i.\end{cases} \tag{2}$$

Let us verify the solution for $p \neq {}^1/_2$, by computing $N(I-Q)$. Its i,j-th entry is

$$\frac{1}{(p-q)(r^n-1)}\left[\sum_{k=1}^{i}(r^k-1)(r^{n-i}-1)s_{kj} + \sum_{k=i+1}^{n-1}(r^i-1)(r^{n-i}-r^{k-i})s_{kj}\right].$$

We recall that $s_{kj}\neq 0$ only for $k=j-1, j, j+1$. Suppose that $j<i$. Then all terms in the second sum are 0. The first sum simplifies to

$$\frac{r^{n-i}-1}{(p-q)(r^n-1)}\left[(r^{j-1}-1)(-p)+(r^j-1)(1)+(r^{j+1}-1)(-q)\right]$$
$$= \frac{r^{n-i}-1}{(p-q)(r^n-1)}\left[r^j\left(-\frac{p}{r}+1-rq\right)+(p-1+q)\right] = 0.$$

The answer for $j > i$ is also 0. Hence all off-diagonal entries are 0. Let us compute the three non-zero terms for the i,i-th entry.

$$\frac{1}{(p-q)(r^n-1)}[(r^{i-1}-1)(r^{n-i}-1)(-p)+(r^i-1)(r^{n-i}-1)(1) + (r^i-1)(r^{n-i}-r)(-q)]$$

$$= \frac{1}{(p-q)(r^n-1)}\left[r^n\left(-\frac{p}{r}+1-q\right)+r^{n-i}(p-1+q) + r^i\left(\frac{p}{r}-1+rq\right)+(-p+1-rq)\right]$$

$$= \frac{r^n(p-q)+(q-p)}{(p-q)(r^n-1)} = 1.$$

Thus $N = (I-Q)^{-1}$ as required.

The solution for $p = 1/2$ may be verified similarly. We can also obtain it by a limit process from the general solution: We write $p-q$ as $q(r-1)$, and we let $q \to 1/2$, $r \to 1$.

Let us next compute τ.

$$t_i = \sum_{j=1}^{n-1} n_{ij} = \frac{1}{(p-q)(r^n-1)} \left[\sum_{j=1}^{i}(r^j-1)(r^{n-i}-1)+\sum_{j=i+1}^{n-1}(r^i-1)(r^{n-i}-r^{j-i})\right]$$

$$= \frac{1}{(p-q)(r^n-1)}\left[(r^{n-i}-1)\left(\sum_{j=1}^{i} r^j - i\right)+(r^i-1) \left((n-i-1)r^{n-i}-\sum_{j=i+1}^{n-1} r^{j-i}\right)\right]$$

$$= \frac{(n-i)r^n - nr^{n-i}+i}{(p-q)(r^n-1)}.$$

Or more simply,

$$t_i = \frac{1}{p-q}\left(n\,\frac{r^n-r^{n-i}}{r^n-1}-i\right) \quad \text{if } p \neq 1/2. \tag{3}$$

Similarly,

$$t_i = i(n-i) \quad \text{if } p = 1/2. \tag{4}$$

An interesting question to consider is finding the maximum value of t_i. For $p = 1/2$ we see that t_i increases till the middle, and then decreases symmetrically. If n is even, the mid-point $i = n/2$ yields $t_i = (n/2)^2$. For the general case we can write down the ratio of two terms, and find the value of i for which this ratio is one. This will not yield an integer in general, but the i nearest will give a maximum.

We find the approximate solution

$$i_{\max} = \log_r((r-1)n).$$

If $p > q$, that is $r > 1$, then the resulting $t_{\max}$ is of the order of magnitude of $(n - i_{\max})/(p-q)$. Thus we see that if $p \neq {}^1/_2$ the absorption time is of a lower order of magnitude for large n than for $p = {}^1/_2$. For example, if $p = {}^2/_3$, and hence $r = 2$, $i_{\max} = \log_2 n$, and $t_{\max}$ is about $3(n - \log_2 n)$ for large n. For the very small $n = 5$ of our example $i_{\max} = 2.3$, and $i = 2$ is the maximum point, but the value $t_{\max}$ obtained is too large for so small an n.

Finally, we shall compute B. It is sufficient to find b_{in}, since $b_{i0} = 1 - b_{in}$. We note that r_{in} is 0 except for $r_{n-1,n} = p$.

$$b_{in} = \sum_{k=1}^{n-1} n_{ik} r_{kn} = n_{i,n-1} \cdot p = \frac{p(r^i - 1)(r^{n-i} - r^{n-1-i})}{(p-q)(r^n - 1)}.$$

Hence

$$b_{in} = \frac{r^n - r^{n-i}}{r^n - 1} \quad \text{if } p \neq {}^1/_2. \tag{5}$$

Similarly,

$$b_{in} = i/n \quad \text{if } p = {}^1/_2. \tag{6}$$

For $p = {}^1/_2$ the solution is very intuitive. The probability of ending up at the right-hand boundary is proportional to the original distance from the left-hand boundary. But the solution for $p \neq {}^1/_2$ has some surprising features. Let us study the case $p > {}^1/_2$, that is $r > 1$. We find that for a given starting position i the probability of ending up at n is not negligible for any n. Indeed, if we keep i fixed and let n tend to infinity, b_{in} approaches the limit $1 - r^{-i}$. If i is fairly large this probability will be close to 1 no matter how large n is! Even for $i = 1$ we have a probability $(r-1)/r$ of ending up at n. The absorption time in this case is about n/p, surprisingly small. For $p = {}^2/_3$ this probability is ${}^1/_2$. This means that if $p = {}^2/_3$ we may put the right-hand boundary n as far out as we wish, start the process at $i = 1$, and still have a better than even chance of ending up at n rather than at 0.

This particular random walk is often referred to as "gambler's ruin:" We may think of two men playing a certain game repeatedly in which player A has probability p of winning. Let i dollars be his original capital, $n - i$ the fortune of his opponent, and assume that 1 dollar is bet each time. Then A's fortune carries out the random walk AARW with the given p. Absorption at n means that A ends up with all the money, while absorption at 0 means that he is ruined. We see that for a fair game ($p = {}^1/_2$) the probability of ruin is equal to the fraction of the two fortunes held by the opponent. But the nature of the solution changes drastically if player A has an advantage in the game. In this case he has a good chance of winning out even if his opponent has a much greater capital. For example, if he has $p = {}^2/_3$, that is $r = 2$ (meaning the odds are 2:1 in his favor), then he has a better than even chance of ruining a rich opponent even if he has only 1 dollar to start with!

We will briefly mention two applications of this result. First of all, gambling houses can exist due to it. They fix the odds so that $r > 1$. Then by making sure that their original capital i is large enough (measured in terms of the size of one wager), they will have probability $1 - r^{-i}$, very near to 1, of staying in business no matter how much is bet at their gambling tables. We also see that the absorption time is enormous, which is the reason that gambling houses have not yet acquired all the money in the world. For r near to 1, (3) is roughly equal to (4). We may estimate, very conservatively, that the gambling house can cover 10,000 bets, while the gamblers can provide 1,000,000 bets. Then $i(n-i)$ is about 10^{10}; which would put the absorption time into thousands of years. This leaves ample opportunity for the raising of new gamblers.

A second application is to a simple model for the principle of natural selection in the theory of evolution. Suppose that on an isolated island the population of some species is fixed at n by the supply of food. Let us suppose that a mutant is born with a slightly better chance of survival than the regular member of the species. A simple model of the struggle for survival is given by assuming that in each generation the mutants gain one place with probability $p > {}^1/_2$ or lose one place with probability q. We then know that the mutants have probability of more than $(r-1)/r$ of taking over the island. If $p = .51$, this probability is .04; if $p = .6$, the probability is ${}^1/_3$. Hence we see that relatively minor advantages can result in the survival of the mutants.

While this simple model serves to illustrate how mutants may take over a large species, the estimate for the absorption time is unrealistic. Even for $p = .6$ and n as small as 100 we obtain n/p or about 167 generations before the mutants take over. This brings out the unrealistic nature of the assumption that only one place is changed in each generation. For a realistic time for absorption we need a more sophisticated model.

We will now show that the solutions for the other three random walks may be obtained from AARW by various tricks. Let us first illustrate APRW for $n = 5$.

APRW for $n=5$, $p={}^1/_2$

$$
P = \begin{array}{c} \\ 0 \\ 1 \\ 2 \\ 3 \\ 4 \\ 5 \end{array}
\begin{array}{c|ccccc}
0 & 1 & 2 & 3 & 4 & 5 \\
1 & 0 & 0 & 0 & 0 & 0 \\ \hline
1/2 & 0 & 1/2 & 0 & 0 & 0 \\
0 & 1/2 & 0 & 1/2 & 0 & 0 \\
0 & 0 & 1/2 & 0 & 1/2 & 0 \\
0 & 0 & 0 & 1/2 & 0 & 1/2 \\
0 & 0 & 0 & 0 & 1/2 & 1/2
\end{array}
$$

$$
N = \begin{array}{c} 1 \\ 2 \\ 3 \\ 4 \\ 5 \end{array}
\begin{pmatrix}
2 & 2 & 2 & 2 & 2 \\
2 & 4 & 4 & 4 & 4 \\
2 & 4 & 6 & 6 & 6 \\
2 & 4 & 6 & 8 & 8 \\
2 & 4 & 6 & 8 & 10
\end{pmatrix}
\qquad
\tau = \begin{pmatrix} 10 \\ 18 \\ 24 \\ 28 \\ 30 \end{pmatrix}
$$

APRW for $n=5$, $p={}^2/_3$

$$
P = \begin{array}{c} \\ 0 \\ 1 \\ 2 \\ 3 \\ 4 \\ 5 \end{array}
\begin{array}{c|ccccc}
0 & 1 & 2 & 3 & 4 & 5 \\
1 & 0 & 0 & 0 & 0 & 0 \\ \hline
1/3 & 0 & 2/3 & 0 & 0 & 0 \\
0 & 1/3 & 0 & 2/3 & 0 & 0 \\
0 & 0 & 1/3 & 0 & 2/3 & 0 \\
0 & 0 & 0 & 1/3 & 0 & 2/3 \\
0 & 0 & 0 & 0 & 1/3 & 2/3
\end{array}
$$

$$
N = \begin{array}{c} 1 \\ 2 \\ 3 \\ 4 \\ 5 \end{array}
\begin{pmatrix}
3 & 6 & 12 & 24 & 48 \\
3 & 9 & 18 & 36 & 72 \\
3 & 9 & 21 & 42 & 84 \\
3 & 9 & 21 & 45 & 90 \\
3 & 9 & 21 & 45 & 93
\end{pmatrix}
\qquad
\tau = \begin{pmatrix} 93 \\ 138 \\ 159 \\ 168 \\ 171 \end{pmatrix}.
$$

The N matrix for APRW may be obtained from AARW through the following observations. Consider $i>j$. If the process

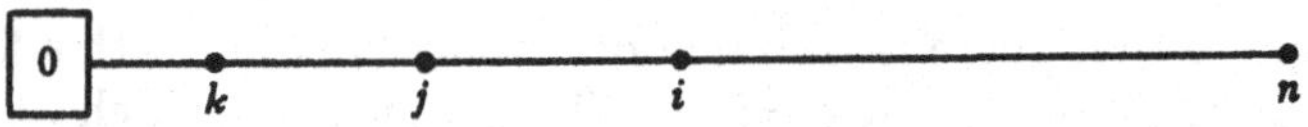

FIGURE 7-3

starts in i, it must eventually enter j. Hence $h_{ij} = n_{ij}/n_{jj} = 1$, and $n_{ij} = n_{jj}$. (This is conspicuous in N above.) If $k<j$, then h_{kj} may be gotten as the probability of ending up at j in AARW with $n=j$. Hence $n_{kj} = h_{kj}n_{jj} = b_{kj}n_{jj}$, with b_{kj} from AARW. Thus it suffices to find n_{jj}. Let us first compute the case $p = 1/2$. We note that Q (except for the last row and column) is a submatrix of Q for larger n. And in the larger Q these rows and columns are filled out with 0's. Hence n_{ij} is independent of n. But as we let $n \to \infty$ in AARW, the difference between it and APRW disappears. Hence n_{jj} for APRW is the limit as $n \to \infty$ of n_{jj} for AARW. Thus

$$\left.\begin{aligned} n_{ij} &= 2j && \text{if } i \geqslant j \\ n_{ij} &= \frac{i}{j}\cdot 2j = 2i && \text{if } i \leqslant j \end{aligned}\right\} \text{ for } p = 1/2. \tag{7}$$

The same argument is applicable if $r < 1$. Thus

$$\left.\begin{aligned} n_{ij} &= \frac{r^j - 1}{p-q} && \text{if } i \geqslant j \\ n_{ij} &= \frac{r^j - r^{j-i}}{r^j - 1}\cdot\frac{r^j - 1}{p-q} = \frac{r^j - r^{j-i}}{p-q} && \text{if } i \leqslant j \end{aligned}\right\} \text{ for } p \neq 1/2. \tag{8}$$

We can also obtain n_{ij} for $i \neq j$ from N for AARW by letting $n \to \infty$.

For $r > 1$ the above argument breaks down, since no matter how large n is in AARW, the probability of ending up at n does not become negligible. But here we make use of the fact that if in AARW we renumber state i as $n-i$, we have the original process with p and q interchanged. We then find that the above formulas also hold for $r > 1$.

For τ we obtain

$$t_i = (2n - i + 1)i \qquad \text{if } p = 1/2; \tag{9}$$

$$t_i = \frac{1}{p-q}\cdot\left[\frac{r^{n+1} - r^{n-i+1}}{r-1} - i\right] \qquad \text{if } p \neq 1/2. \tag{10}$$

Since there is only one absorbing state, all absorption probabilities are 1.

In both AARW and APRW there is a simple relation between n_{ij} and n_{ji}. In fact it may be verified from our formulas for these quantities that

$$n_{ij} = r^{j-i} n_{ji}.$$

We shall see that there is a simple probabilistic proof of this fact.

Assume that $j < i$. Let $d = i - j$. Then any path which allows the process to go from i to j in n steps must take d more steps to the left than to the right. Hence the probability that the process starting at i will follow such a path is

$$p^k q^{k+d}$$

where $2k+d=n$. On the other hand each such path, looked at backwards, may be considered a path from j to i. For the process to follow this path it must make d more steps to the right than to the left. Hence the probability that, starting at j, the process will follow this path is

$$p^{k+d} q^k.$$

There are the same number of paths from i to j in n steps as there are from j to i, but the ratio of the probability for each path is

$$\frac{p^k q^{k+d}}{p^{k+d} q^k} = \frac{q^d}{p^d} = r^{j-i}.$$

Hence, $p^{(n)}{}_{ij} = r^{j-i} p^{(n)}{}_{ji}$. But then

$$n_{ij} = \sum_{n=0}^{\infty} p^{(n)}{}_{ij} = r^{j-i} \sum_{n=0}^{\infty} p^{(n)}{}_{ji} = r^{j-i} n_{ji}.$$

Note that this also shows that when $p=1/2$, then $r=1$, $n_{ij}=n_{ji}$.

We now turn to the regular chains, starting with PPRW.

PPRW for $n=5$, $p=1/2$

$$P = \begin{array}{c} \\ 0 \\ 1 \\ 2 \\ 3 \\ 4 \\ 5 \end{array} \begin{array}{c} \begin{array}{cccccc} 0 & 1 & 2 & 3 & 4 & 5 \end{array} \\ \begin{pmatrix} 1/2 & 1/2 & 0 & 0 & 0 & 0 \\ 1/2 & 0 & 1/2 & 0 & 0 & 0 \\ 0 & 1/2 & 0 & 1/2 & 0 & 0 \\ 0 & 0 & 1/2 & 0 & 1/2 & 0 \\ 0 & 0 & 0 & 1/2 & 0 & 1/2 \\ 0 & 0 & 0 & 0 & 1/2 & 1/2 \end{pmatrix} \end{array}$$

$$\alpha = (1/6, 1/6, 1/6, 1/6, 1/6, 1/6)$$

$$M = \begin{array}{c} \\ 0 \\ 1 \\ 2 \\ 3 \\ 4 \\ 5 \end{array} \begin{array}{c} \begin{array}{cccccc} 0 & 1 & 2 & 3 & 4 & 5 \end{array} \\ \begin{pmatrix} 6 & 2 & 6 & 12 & 20 & 30 \\ 10 & 6 & 4 & 10 & 18 & 28 \\ 18 & 8 & 6 & 6 & 14 & 24 \\ 24 & 14 & 6 & 6 & 8 & 18 \\ 28 & 18 & 10 & 4 & 6 & 10 \\ 30 & 20 & 12 & 6 & 2 & 6 \end{pmatrix} \end{array}$$

PPRW for $n=5$, $p=2/3$

$$P = \begin{array}{c} \\ 0 \\ 1 \\ 2 \\ 3 \\ 4 \\ 5 \end{array}\begin{array}{c} \begin{array}{cccccc} 0 & 1 & 2 & 3 & 4 & 5 \end{array} \\ \begin{pmatrix} 1/3 & 2/3 & 0 & 0 & 0 & 0 \\ 1/3 & 0 & 2/3 & 0 & 0 & 0 \\ 0 & 1/3 & 0 & 2/3 & 0 & 0 \\ 0 & 0 & 1/3 & 0 & 2/3 & 0 \\ 0 & 0 & 0 & 1/3 & 0 & 2/3 \\ 0 & 0 & 0 & 0 & 1/3 & 2/3 \end{pmatrix} \end{array}$$

$$\alpha = (1/63, 2/63, 4/63, 8/63, 16/63, 32/63)$$

$$M = \begin{array}{c} \\ 0 \\ 1 \\ 2 \\ 3 \\ 4 \\ 5 \end{array}\begin{array}{c} \begin{array}{cccccc} 0 & 1 & 2 & 3 & 4 & 5 \end{array} \\ \begin{pmatrix} 63 & 3/2 & 15/4 & 51/8 & 147/16 & 387/32 \\ 93 & 63/2 & 9/4 & 39/8 & 123/16 & 339/32 \\ 138 & 45 & 63/4 & 21/8 & 87/16 & 267/32 \\ 159 & 66 & 21 & 63/8 & 45/16 & 183/32 \\ 168 & 75 & 30 & 9 & 63/16 & 93/32 \\ 171 & 78 & 33 & 12 & 3 & 63/32 \end{pmatrix} \end{array}$$

We may obtain APRW from PPRW by making 0 absorbing. Thus, if we know α for PPRW, we may obtain M from N of APRW by Corollary 6.2.6,

$$m_{ij} = \frac{1}{a_j}(n_{jj} - n_{ij}) + t_i - t_j \quad \text{for } i \neq j.$$

If $p=1/2$, we have column sums 1, and hence $\alpha = \frac{1}{n+1}\eta$. Thus we obtain

$$m_{ij} = \begin{cases} n+1 & i = j \\ (2n-i+1)i-(2n-j+1)j & i > j \\ j(j+1)-i(i+1) & i < j \end{cases} \quad p = 1/2. \tag{11}$$

For $p \neq 1/2$ the above example suggests that $a_{i+1} = ra_i$. Indeed, this yields a fixed vector, as can be seen from writing our equations as

$$qa_0 + qa_1 = a_0 \qquad \text{or} \quad a_1 = ra_0$$

$$pa_{i-1} + qa_{i+1} = a_i \quad \text{or} \quad \frac{p}{r}a_{i+1} + qra_{i-1} = a_i$$

$$pa_{n-1} + pa_n = a_n \qquad \text{or} \quad a_{n-1} = \frac{1}{r}a_n.$$

Thus

$$a_i = \frac{r-1}{r^{n+1}-1}\, r^i.$$

Then

$$m_{ij} = \left\{\begin{array}{ll} \dfrac{r^{n+1}-1}{r-1}\cdot\dfrac{1}{r^i} & i = j \\[2ex] \dfrac{1}{p-q}\cdot\left[\dfrac{r^{n-j+1}-r^{n-i+1}}{r-1} - (i-j)\right] & i > j \\[2ex] \dfrac{1}{p-q}\cdot\left[(j-i - \dfrac{r^j-r^i}{(r-1)r^{i+j}}\right] & i < j \end{array}\right\}\quad p \neq {}^1/_2. \tag{12}$$

Let us compare m_{0n} with m_{n0}. If $p={}^1/_2$, $m_{0n}=m_{n0}=n(n+1)$, which is of the order n^2 for large n. But if $p \neq {}^1/_2$, we observe a completely different behavior. Say $r>1$, then

$$m_{0n} = \frac{1}{p-q}\cdot\left[n - \frac{r^n-1}{(r-1)r^n}\right]$$

is roughly $\dfrac{1}{p-q}\cdot n$ for large n. Hence it takes a surprisingly short time to go "all the way" in the favored direction.

$$m_{n0} = \frac{1}{p-q}\cdot\left[\frac{r^{n+1}-r}{r-1} - n\right]$$

which is roughly $\dfrac{r}{(p-q)(r-1)}\cdot r^n$ for large n. Since $r>1$, this increases exponentially in n. For very large n it will take a tremendously long time to go "all the way" in the wrong direction. This type of behavior is typical of random walks. We will see another example of this in the Ehrenfest model.

Finally we consider CRW.

CRW for $n=5$, $p={}^1/_2$

$$P = \begin{array}{c} \\ 1\\2\\3\\4\\5 \end{array}\!\!\begin{array}{c} \begin{array}{ccccc} 1 & 2 & 3 & 4 & 5 \end{array} \\ \begin{pmatrix} 0 & {}^1/_2 & 0 & 0 & {}^1/_2 \\ {}^1/_2 & 0 & {}^1/_2 & 0 & 0 \\ 0 & {}^1/_2 & 0 & {}^1/_2 & 0 \\ 0 & 0 & {}^1/_2 & 0 & {}^1/_2 \\ {}^1/_2 & 0 & 0 & {}^1/_2 & 0 \end{pmatrix} \end{array}$$

$$M = \begin{array}{c} 1\\2\\3\\4\\5 \end{array}\!\!\begin{pmatrix} 5 & 4 & 6 & 6 & 4 \\ 4 & 5 & 4 & 6 & 6 \\ 6 & 4 & 5 & 4 & 6 \\ 6 & 6 & 4 & 5 & 4 \\ 4 & 6 & 6 & 4 & 5 \end{pmatrix}$$

CRW for $n=5$, $p={}^2/_3$

$$P = \begin{array}{c} \\ 1 \\ 2 \\ 3 \\ 4 \\ 5 \end{array}\begin{array}{c} \begin{array}{ccccc} 1 & 2 & 3 & 4 & 5 \end{array} \\ \begin{pmatrix} 0 & {}^2/_3 & 0 & 0 & {}^1/_3 \\ {}^1/_3 & 0 & {}^2/_3 & 0 & 0 \\ 0 & {}^1/_3 & 0 & {}^2/_3 & 0 \\ 0 & 0 & {}^1/_3 & 0 & {}^2/_3 \\ {}^2/_3 & 0 & 0 & {}^1/_3 & 0 \end{pmatrix} \end{array}$$

$$M = \begin{array}{c} 1 \\ 2 \\ 3 \\ 4 \\ 5 \end{array}\begin{pmatrix} 5 & {}^{78}/_{31} & {}^{141}/_{31} & {}^{174}/_{31} & {}^{147}/_{31} \\ {}^{147}/_{31} & 5 & {}^{78}/_{31} & {}^{141}/_{31} & {}^{174}/_{31} \\ {}^{174}/_{31} & {}^{147}/_{31} & 5 & {}^{78}/_{31} & {}^{141}/_{31} \\ {}^{141}/_{31} & {}^{174}/_{31} & {}^{147}/_{31} & 5 & {}^{78}/_{31} \\ {}^{78}/_{31} & {}^{141}/_{31} & {}^{174}/_{31} & {}^{147}/_{31} & 5 \end{pmatrix}$$

This process always has $\alpha=\frac{1}{n}\eta$, since P always has column sums 1. This is also obvious from the fact that no position in the circle is distinguished. For the same reason we expect that m_{ij} should depend only on how far i and j are apart (and on the direction if $p\neq{}^1/_2$). This is certainly so in the examples above.

If state n is made absorbing in CRW, we obtain AARW—with 0 and n identified. Hence M for CRW is obtainable from N for AARW. But there is an even simpler method. If we want m_{ij}, we renumber the states so that j becomes n, and then m_{ij} is just an absorption time for AARW. Specifically, if the distance from j to i (clockwise) is d, then $m_{ij}=t_d$. Thus

$$\begin{aligned} m_{ii} &= n \\ m_{ij} &= \frac{1}{p-q}\cdot\left\{n\,\frac{r^n-r^{n-d}}{r^n-1}-d\right\} \qquad p\neq{}^1/_2,\ i\neq j \qquad (13) \\ m_{ij} &= d(n-d) \qquad\qquad p={}^1/_2,\ i\neq j \end{aligned}$$

where d is the clockwise distance from j to i.

The remarks made for τ of AARW are applicable here. Thus, for example, the distance (clockwise) that it takes longest to travel is approximately $\log_r((r-1)n)$. We also note that m_{ij} is generally of a lower order of magnitude for any $p\neq{}^1/_2$ than for $p={}^1/_2$.

Let us next find the transition matrices for the reverse processes of PPRW and CRW.

For PPRW, remembering that $a_{i+1} = ra_i$,

$$\hat{p}_{i,i+1} = \frac{a_{i+1}}{a_i} p_{i+1,i} = rq = p = p_{i,i+1}$$

$$\hat{p}_{i+1,i} = \frac{a_i}{a_{i+1}} p_{i,i+1} = \frac{1}{r} p = q = p_{i+1,i}.$$

Hence the process is reversible, contrary to one's intuition. But for CRW $a_{i+1} = a_i$, hence

$$\hat{p}_{i,i+1} = \frac{a_{i+1}}{a_i} p_{i+1,i} = q$$

$$\hat{p}_{i+1,i} = \frac{a_i}{a_{i+1}} p_{i,i+1} = p.$$

Hence the reverse process is a CRW with p replaced by q, as one would expect. It is reversible only when $p = q = {}^1/_2$.

Let us close by giving a practical application for PPRW. We consider a gambling house that wishes to keep a close check on its roulette wheel. Suppose that it is a wheel having in addition to the numbers 1 to 36, half of which are red and half black, the numbers 0 and 00 which are not colored. We will devise a simple automatic check to see that the house is taking its share of the bets on red. We set up an electric counter which starts at 0 and adds 1 every time red comes up, subtracts 1 every time red fails to come up. The counter does not go below 0. If the counter reaches a specified number n, then it rings a bell, and the house changes the wheel.

If the wheel is properly balanced, then we have PPRW with $p = {}^{18}/_{38} = {}^9/_{19}$. Hence it will take a very long time to reach n. The house adjusts n so that m_{0n} corresponds to its normal periodic servicing of the wheel. However, if the wheel fails to function properly—for example, if p rises to ${}^1/_2$, in which case the house no longer makes a profit—then the bell will ring much sooner. A similar check on black will assure that the house continues to make its profit on all bets.

Let us consider a concrete example of this. Suppose that $n = 40$ is selected. Then m_{0n} for proper functioning is about 18,000. If the wheel is turned 400 times in a day, the bell will ring on the average once in 45 days, allowing for normal servicing. But if p rises to ${}^1/_2$, then $m_{0n} = 1640$, and the bell will ring after four days. If p rises above the break-even figure (that is, the house is losing money), then the bell will ring very quickly.

§ 7.2 Applications to sports. Let us apply some of our results to the game of tennis. We will first consider the problem of a single game of tennis played between two players. We will assume that player A

has probability p of winning any given point, and player B has probability $q \leqslant p$.

If we keep score in the ordinary manner, there are 20 possible scores during a game. These are: 0–0, 0–15, 15–0, 0–30, 15–15, 30–0, 0–40, 15–30, 30–15, 40–0, 15–40, 30–30, 40–15, 30–40, 40–30, Advantage B, Deuce, Advantage A, Game B, Game A. However, it is easily seen

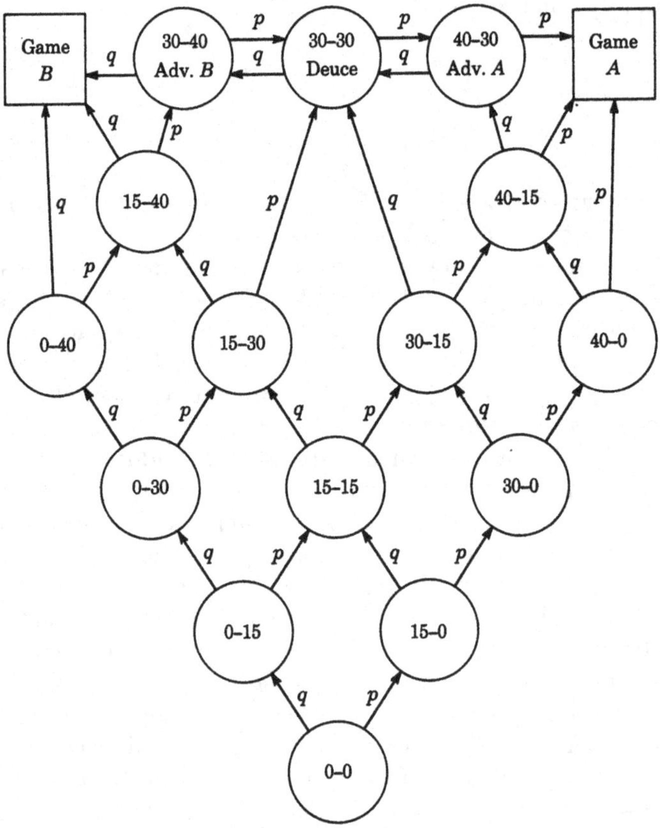

FIGURE 7-4

that we may lump the following pairs: {30–30, Deuce}, {30–40, Advantage B}, {40–30, Advantage A}. The resulting random walk is then represented by Figure 7-4.

The game of tennis may conveniently be broken down into two stages. At the beginning the process goes through some of the lower twelve states, always moving up, and in four or five steps it arrives at one of the five states in the top row. This we will refer to as the *preliminary process.* The preliminary process is of an extremely simple nature. It is followed by a random walk of type AARW, with $n=4$. The state Game B is the absorbing state 0 of AARW, and Game A is the absorbing state n.

We will describe the entire process as an AARW random walk, with initial probabilities furnished by the preliminary process. These initial probabilities $\pi=(c_0, c_1, c_2, c_3, c_4)$ are given by an elementary probability calculation. We find that

$$c_0 = q^4(1+4p), \quad c_1 = 4p^2q^3, \quad c_2 = 6p^2q^2,$$
$$c_3 = 4p^3q^2, \quad c_4 = p^4(1+4q).$$

If $p={}^1/_2$, $c_0={}^3/_{16}$, $c_1={}^1/_8$, $c_2={}^3/_8$, $c_3={}^1/_8$, $c_4={}^3/_{16}$. Using the basic quantities of AARW with $n=4$,

$$N = \frac{1}{(p-q)(r^4-1)}\begin{pmatrix}(r-1)(r^3-1) & (r-1)(r^3-r) & (r-1)(r^3-r^2)\\ (r-1)(r^2-1) & (r^2-1)^2 & (r^2-1)(r^2-r)\\ (r-1)^2 & (r-1)(r^2-1) & (r-1)((r^3-1)\end{pmatrix}$$

$$\tau = \frac{1}{p-q}\begin{pmatrix}4\,\dfrac{r^4-r^3}{r^4-1}-1\\[2ex] 4\,\dfrac{r^4-r^2}{r^4-1}-2\\[2ex] 4\,\dfrac{r^4-r}{r^4-1}-3\end{pmatrix} \qquad \{b_{i4}\} = \frac{1}{r^4-1}\begin{pmatrix}r^4-r^3\\ r^4-r^2\\ r^4-r\end{pmatrix} \qquad p \neq {}^1/_2$$

$$N = \begin{pmatrix}{}^3/_2 & 1 & {}^1/_2\\ 1 & 2 & 1\\ {}^1/_2 & 1 & {}^3/_2\end{pmatrix} \qquad \xi = \begin{pmatrix}3\\ 4\\ 3\end{pmatrix} \qquad \{b_{i4}\} = \begin{pmatrix}{}^1/_4\\ {}^1/_2\\ {}^3/_4\end{pmatrix} \qquad p = {}^1/_2$$

we can find all interesting quantities. The most interesting one is, of course, the probability that A will win. For $p={}^1/_2$ we obtain

$$({}^3/_{16})\cdot 0 + ({}^1/_8)\cdot({}^1/_4) + ({}^3/_8)\cdot({}^1/_2) + ({}^1/_8)\cdot({}^3/_4) + ({}^3/_{16})\cdot 1 = {}^1/_2.$$

This was to be expected, by symmetry. If $p>{}^1/_2$, we obtain

$$\frac{1}{r^4-1}\,[q^4(1+4p)\cdot 0 + 4p^2q^3(r^4-r^3) + 6p^2q^2(r^4-r^2) + 4p^3q^2(r^4-r)$$
$$+ p^4(1+4q)(r^4-1)]$$

which simplifies to

$$p_A = \frac{p^4(1-16q^4)}{p^4-q^4}.$$

For example, if $p = .51$, then $p_A = .525$, and if $p = .6$, then $p_A = .736$. For the absorption time and the number of times in a state we will carry out the computation only when $p = 1/2$. The interesting cases are ones where p is near $1/2$, and absorption times do not depend very drastically on p.

When $p = 1/2$ we find that the mean number of times in the three interesting transient states is: 1 for Deuce, and $5/8$ for Advantage B and for Advantage A. The absorption time is $9/4$. To find the actual length of the game, we must also take into account the preliminary stage. If we do, we find that the mean length of a tennis game between equally matched opponents is $27/4 = 6\,3/4$. Since the minimum length of the game is 4, this shows that for equally matched players the average length is not much above the minimum.

Should it prove from records that games actually are much longer than this, and that the average number of times in Deuce is well above 1, as seems to be the case, then it would indicate that the present model for tennis is too simple. Perhaps a player "plays harder" when he is behind. This would lead to a somewhat more complicated random walk.

Let us return to the probability that the better player A wins. This is always greater for a game than for an individual point. Thus, the game serves to magnify the difference between the two players. This is further magnified since several games are played in a set, and several sets in a match. The probabilities for a set, in which a player must win at least six games, but by a margin of at least two, may be computed just as above. We are led to the same AARW, but with a longer preliminary stage. A match is won by the first player winning three sets. This is a straightforward computation. The following figures will illustrate the magnification achieved in sets and matches:

	$p = .51$	$p = .6$
Probability of winning point	.510	.600
Probability of winning game	.525	.736
Probability of winning set	.573	.966
Probability of winning match	.635	.9996

Thus there is always a good chance that the better player will win the match. And if there is a fairly significant difference between players, then it is practically certain that the better player wins.

Let us compare these results for tennis with the World Series in baseball. Here the team that first wins four games is declared winner. If we assume that team A has probability p of winning any one game, then we find that

$$p_A = p^4(1 + 4q + 10q^2 + 20q^3),$$

where the various terms correspond to series of 4, 5, 6, and 7 games respectively. If $p = .51$, $p_A = .522$, while if $p = .6$, then $p_A = .710$. The World Series also magnifies differences between teams, but not nearly as well as a match of tennis—or even a single game of tennis.

If we compute the mean length of a series, we find this to be $t = 4(p^4 + q^4) + 20(p^4q + pq^4) + 60(p^4q^2 + p^2q^4) + 140(p^4q^3 + p^3q^4)$. This is largest (5.81) when $p = {}^1/_2$, and decreases monotonically to 4 as p is increased to 1. Hence we should be able to estimate p from the observed length of World Series. In the 50 World Series played under the stated rules from 1905 to 1957, 10 ended in four games, 13 in five games, 12 in six games, and 15 in seven games. This yields a mean length of 5.64, which would agree very well with $p = {}^5/_8$. This would suggest that the teams playing in the World Series have, on the average, not been matched too closely.

Let us now consider the efficiency of various procedures in magnifying the differences between players or teams. We have found that tennis (one game) gives slightly more magnification than the World Series, but it also requires more steps on the average. To be able to compare the efficiency of two rules, we will have to take them so that the mean length of a series is the same.

Tennis may be compared to the World Series as follows: The latter requires four wins, while the former requires that the winner have four wins and be ahead by two. This is really a hybrid between two types of procedures. The pure procedure would be to require that the winner end up ahead by four points. It can easily be seen that this rule gives much more magnification than the other, but also requires a great deal more time. Let us therefore consider two classes of rules.

RULE W_n: *The first person to win n points is declared winner.*

RULE A_n: *The first person to get ahead of his opponent by n points is declared winner.*

We will compare these two rules, selecting n in each case so that the mean length of a game be a given large number, which we will denote by N^2, and seeing how much they magnify differences. Since we will be interested particularly in large n, we will allow asymptotic approximations. Since we are particularly interested in magnification of very small differences, we will let the players differ by ϵ, that is let $p = (1 + \epsilon)/2$ and $q = (1 - \epsilon)/2$, and compute the final difference just to the first order term in ϵ.

The following identities will be useful:

$$\sum_{k=0}^{n} \binom{n+k}{k} = \binom{2n+1}{n}$$

$$\sum_{k=0}^{n} \binom{n+k}{k} k = n\binom{2n+1}{n} - \binom{2n+1}{n-1}$$

$$\sum_{k=0}^{n} \binom{n+k}{k}\left(\frac{1}{2}\right)^k = 2^n$$

$$\sum_{k=0}^{n} \binom{n+k}{k}\left(\frac{1}{2}\right)^k \cdot k = (n+1)2^n - \frac{2n+1}{2^n}\binom{2n}{n}$$

If we use W_n, the probability that the better player wins is

$$\begin{aligned} p_A &= p^n \sum_{k=0}^{n-1} \binom{n-1+k}{k} q^k \\ &= \left(\frac{1+\epsilon}{2}\right)^n \sum_{k=0}^{n-1} \binom{n-1+k}{k}\left(\frac{1-\epsilon}{2}\right)^k \\ &\approx \frac{1+n\epsilon}{2^n} \sum_{k=0}^{n-1} \binom{n-1+k}{k}\left(\left(\frac{1}{2}\right)^k - \epsilon k\left(\frac{1}{2}\right)^k\right) \\ &= \frac{1+n\epsilon}{2^n}\left(2^{n-1} - \epsilon\left[n2^{n-1} - \frac{2n-1}{2^{n-1}}\binom{2n-2}{n-1}\right]\right) \end{aligned}$$

Using Stirling's formula† and simplifying we obtain

$$p_A \approx \frac{1}{2} + \sqrt{\frac{n}{\pi}}\,\epsilon$$

The expected length of the game (for which we may use $p = 1/2$) is

$$\begin{aligned} t &= 2\cdot(1/2)^n \sum_{k=0}^{n-1} \binom{n-1+k}{k}(n+k)(1/2)^k \\ &= (1/2)^{n-1}\left[n2^{n-1} + n2^{n-1} - \frac{2n-1}{2^{n-1}}\binom{2n-2}{n-1}\right] \\ &\approx 2n - \frac{2}{\sqrt{\pi}}\sqrt{n}. \end{aligned}$$

We will simply use $t \approx 2n$. Thus if we want $t \approx N^2$, we must choose $n = \frac{N^2}{2}$. This yields $p_A \approx 1/2 + \frac{N}{\sqrt{2\pi}}\,\epsilon$, $p_A - p_B \approx N\sqrt{\frac{2}{\pi}}\,\epsilon$; and thus a magnification factor is $N\sqrt{\frac{2}{\pi}}$ or about $.8N$.

† See W. Feller, *Introduction to Probability Theory and Its Applications*, John Wiley & Son, Inc., New York, 1957, Chapter 2.

The method A_n may be represented as AARW from 0 to $2n$, starting at n. Hence

$$p_A = \frac{r^{2n} - r^n}{r^{2n} - 1}.$$

Here we find that the terms in ϵ cancel and that ϵ^2 terms must be carried. We find

$$r^n \approx 1 + 2n\epsilon + 2(n + n^2)\epsilon^2$$

$$r^{2n} \approx 1 + 4n\epsilon + 4(n + 2n^2)\epsilon^2$$

$$p_A \approx \frac{2n\epsilon + 2(n + 3n^2)\epsilon^2}{4n\epsilon + 4(n + 2n^2)\epsilon^2} = \frac{1/2 + 1/2(1 + 3n)\epsilon}{1 + (1 + 2n)\epsilon}$$

$$\approx 1/2 + \frac{n}{2}\,\epsilon.$$

For the mean length we find (using $p \approx 1/2$) $t = n(2n - n) \approx n^2$. Thus we choose $n = N$, and obtain $p_A \approx 1/2 + \frac{N}{2}\,\epsilon$, $p_A - p_B \approx N\epsilon$; yielding a magnification factor of N.

We thus find that $W_{N^2/2}$ and A_N are comparable, in the sense that each yields an approximate mean time of N^2. The former magnifies minute differences by about $.8N$, while the latter multiplies them by N. Thus the A_n rule is more efficient. Any mixture of the two, as in tennis, will lie in between, and hence will also be less efficient than A_n.

§ 7.3 Ehrenfest model for diffusion. There is a simple model for a system of statistical mechanics which is due to T. Ehrenfest. In this model we consider a gas which is contained in a volume that is divided into two regions A and B by a permeable membrane. We assume that the gas has s molecules. At each instant of time a molecule is chosen at random from the set of s molecules and moved from the region that it is in to the other region. We are interested in the way in which the composition of the two regions changes with time. For example, if we start with all the molecules in one region, how long on the average will it be before each regions has half the molecules? Such questions can be answered by using the methods of Markov chains.

We form a Markov chain as follows: We assume first that the molecules are identifiable. We take as states a vector $\gamma = (x_1, x_2, \ldots, x_s)$ where x_j is 1 if the j-th molecule is in region A, and 0 otherwise. Knowing the state tells us the exact composition of A and hence also of B. There are 2^s states. If the process is in state γ, then choosing a molecule at random and moving it to the other region means that we change the state γ to a state δ by simply changing one coordinate of γ. It is clear that from γ there are s states to which the process can move and these transitions occur each with probability $1/s$. It is possible to go from any state γ to any other state δ in a sequence of steps, but it is possible to go from γ to γ only in an even number of steps; and it is possible in two steps. Hence we have an ergodic chain with period 2. It is clear also that we can go from γ to δ in one step if and only if we go from δ to γ in one step. If possible, the probability in each case is $1/s$. Hence the transition matrix is symmetric. This tells us two things. First, the process is a reversible process. Secondly, the fixed probability vector is a constant vector with components $1/2^s$.

The transition matrix for the case $s=3$ is

$$\begin{array}{c} \\ 000 \\ 001 \\ 010 \\ 100 \\ 011 \\ 101 \\ 110 \\ 111 \end{array}\begin{array}{c} \begin{array}{cccccccc} 000 & 001 & 010 & 100 & 011 & 101 & 110 & 111 \end{array} \\ \begin{pmatrix} 0 & 1/3 & 1/3 & 1/3 & 0 & 0 & 0 & 0 \\ 1/3 & 0 & 0 & 0 & 1/3 & 1/3 & 0 & 0 \\ 1/3 & 0 & 0 & 0 & 1/3 & 0 & 1/3 & 0 \\ 1/3 & 0 & 0 & 0 & 0 & 1/3 & 1/3 & 0 \\ 0 & 1/3 & 1/3 & 0 & 0 & 0 & 0 & 1/3 \\ 0 & 1/3 & 0 & 1/3 & 0 & 0 & 0 & 1/3 \\ 0 & 0 & 1/3 & 1/3 & 0 & 0 & 0 & 1/3 \\ 0 & 0 & 0 & 0 & 1/3 & 1/3 & 1/3 & 0 \end{pmatrix} \end{array}$$

The fixed vector is $\alpha=(1/8, 1/8, 1/8, 1/8, 1/8, 1/8, 1/8, 1/8)$. From this we see, for example, that the mean number of steps required to return to any one state is 8. We also have, by Theorem 6.2.3, that the mean number of times in each of the states between occurrences of a particular state is 1. In general, for a chain with s states the mean time to return to a state will be 2^s and the mean number of times in state δ between occurrences of a state γ is 1.

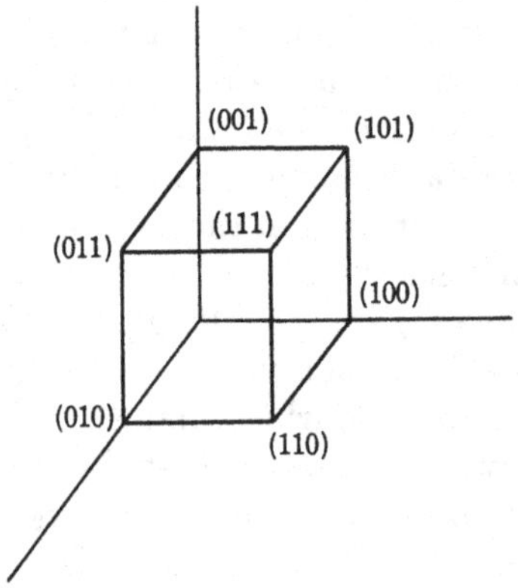

FIGURE 7-5

There is a simple random walk interpretation for the process we are considering. The vectors γ are the corner points of an s-dimensional cube. The possible states to which the process can move from γ in one step are the corner points connected to γ by an edge. There are s such points, and the probability of moving to any one is $1/s$. For the case $s=3$, we have the cube in Figure 7-5.

We define the distance between two states γ and δ, denoted by $d(\gamma, \delta)$, to be the minimum number of steps required to go from γ to δ. In terms of the coordinates of γ and δ this is

$$d(\gamma,\delta) = \sum_{i=1}^{s} |x_i - y_i|.$$

It is clear from the random walk interpretation that the mean time to go from γ to δ depends only on the distance between γ and δ. Let $m^s{}_d$ be this mean time for two points a distance d apart. For fixed s, we compute $m_d = m^s{}_d$ as follows. Let γ and δ be two points a distance d apart. This means that they have exactly d coordinates different. On one step there are d choices which will make the process one unit closer to δ, and $s-d$ choices which will make it one unit farther from δ. Hence, considering the possible first steps, we have

$$m_d = 1 + \frac{d}{s} m_{d-1} + \frac{s-d}{s} m_{d+1}, \qquad 0 < d \leqslant s,$$

where we let $m_0 = m_{s+1} = 0$. These equations have a unique solution which may be written as follows. Let

$$Q^s{}_i = \sum_{k=0}^{i} \frac{\binom{s}{k}}{\binom{s-1}{i}}, \qquad i = 0, 1, \ldots, s-1,$$

then

$$m_d = \sum_{i=1}^{d} Q^s{}_{s-i}, \qquad 0 < d \leqslant s.$$

The values of $Q^s{}_i$ for values of s up to 6 are given in Figure 7-6.

VALUES FOR $Q^s{}_i$

i \ s	1	2	3	4	5	6
0	1	1	1	1	1	1
1		3	2	$1^2/_3$	$1^1/_2$	$1^2/_5$
2			7	$3^2/_3$	$2^2/_3$	$2^1/_5$
3				15	$6^1/_2$	$4^1/_5$
4					31	$11^2/_5$
5						63

FIGURE 7-6

For values of s up to 6 the values of $m^s{}_d$ are given in Figure 7-7.

VALUES FOR $m^s{}_d$

$d \backslash s$	1	2	3	4	5	6
1	1	3	7	15	31	63
2		4	9	$18^2/_3$	$37^1/_2$	$74^2/_5$
3			10	$20^1/_3$	$40^1/_6$	$78^3/_5$
5				$21^1/_3$	$41^2/_3$	$80^4/_5$
5					$42^2/_3$	$82^1/_5$
6						$83^1/_5$

FIGURE 7-7

We note that the means for a given s increase as we increase the distance. This is to be expected. However, they increase very slowly.

We shall call the above chain the *microscopic chain*. In physical applications this chain is often not as interesting as one obtained from it by lumping. The *macroscopic chain* is a Markov chain obtained from the microscopic chain by lumping all states having the same number of molecules in region A. Let us verify that the condition for lumpability is satisfied. Let $\mathbf{V}_i$ be the set of all states in the microscopic process with i molecules in region A. The set $\mathbf{V}_i$ has $\binom{s}{i}$ elements. From any element of $\mathbf{V}_i$ the process moves to one of $\binom{s}{i+1}$ elements of $\mathbf{V}_{i+1}$ or to one of $\binom{s}{i-1}$ elements in $\mathbf{V}_{i-1}$. The probability of moving to $\mathbf{V}_{i+1}$ is $(s-i)/s$ and the probability of moving to $\mathbf{V}_{i-1}$ is i/s. These probabilities are the same for all elements of $\mathbf{V}_i$. Hence the condition for lumpability is satisfied, and we obtain a new Markov chain with states $\mathbf{V}_0, \mathbf{V}_1, \mathbf{V}_2, \ldots, \mathbf{V}_s$. The transition probabilities are

$$
\begin{aligned}
p_{i,i+1} &= 1 - i/s \\
p_{i,i-1} &= i/s \\
p_{i,j} &= 0 \quad \text{otherwise.}
\end{aligned}
$$

We shall refer to this new process as the *macroscopic process*.

By the results of § 6.3 we know that the lumped process will again be ergodic and reversible. It is again of period 2. The fixed vector $\hat{a}$

for the lumped process is easily obtained from α. The component $\hat{a}_i$ is the sum of the components of α for states in $\mathbf{V}_i$. Since there are $\binom{s}{i}$ states in $\mathbf{V}_i$ we have

$$\hat{a}_i = \frac{\binom{s}{i}}{2^s}.$$

Hence

$$\hat{a} = \left\{\frac{\binom{s}{i}}{2^s}\right\}.$$

By Theorem 6.2.3 we see that the mean number of times in any state $\mathbf{V}_i$, between occurrences of state $\mathbf{V}_j$, is

$$\frac{\binom{s}{i}}{\binom{s}{j}}.$$

In particular the mean number of times in each state between occurrences of 0 is $\binom{s}{i}$.

For our example $s = 3$ the partition for the lumping is $\mathbf{V} = (\{000\}, \{100, 010, 001\}, \{110, 101, 011\}, \{111\})$. The transition matrix is

$$\begin{array}{c c} & \begin{array}{cccc} \mathbf{V}_0 & \mathbf{V}_1 & \mathbf{V}_2 & \mathbf{V}_3 \end{array} \\ \begin{array}{c} \mathbf{V}_0 \\ \mathbf{V}_1 \\ \mathbf{V}_2 \\ \mathbf{V}_3 \end{array} & \begin{pmatrix} 0 & 1 & 0 & 0 \\ 1/3 & 0 & 2/3 & 0 \\ 0 & 2/3 & 0 & 1/3 \\ 0 & 0 & 1 & 0 \end{pmatrix} \end{array}.$$

For the macroscopic chain we are also primarily interested in the mean first passage times. We know that in general it is not easy to obtain the mean first passage times for the lumped chain from the original chain. However, in the case we are considering we are helped by two special features of the process. First, for the lumped chain we can obtain all of the values of m_{ij} from the knowledge of only m_{i0}. In fact, since it is possible to go from $\mathbf{V}_{i+1}$ to $\mathbf{V}_0$ only by going through $\mathbf{V}_i$ we have

$$m_{i+1,i} = m_{i+1,0} - m_{i,0} \qquad 0 \leqslant i < s$$

and, by symmetry,

$$m_{i,i+1} = m_{s-i,s-i-1} = m_{s-i,0} - m_{s-i-1,0} \qquad 0 \leqslant i < s.$$

Thus we need only find the mean times to go from any state to $\mathbf{V}_0$. In lumping, the set which became state $\mathbf{V}_0$ was a set with only one element, namely the vector $\gamma = (0, 0, 0, \ldots, 0)$. Hence the mean time to go from state $\mathbf{V}_i$ in the lumped chain to $\mathbf{V}_0$ is the same as the mean time to go from any element in the set $\mathbf{V}_i$ to $\gamma = (0, 0, 0, \ldots, 0)$. But any such element is a distance i from this vector. Hence this mean time is $m^s{}_i$. Using our results for the microscopic process we have the values m_{ij} for the macroscopic process. These are best expressed as follows:

$$m_{i,i+1} = m_{s-i,0} - m_{s-i-1,0} = \sum_{k=1}^{s-i} Q^s{}_{s-k} - \sum_{k=1}^{s-i-1} Q^s{}_{s-k} = Q^s{}_i$$
$$0 \leqslant i < s-1.$$

To go from $\mathbf{V}_i$ to $\mathbf{V}_j$ we must go through every intermediate point. Hence

$$m_{ij} = \sum_{k=i}^{j-1} Q^s{}_k \quad \text{for } i < j$$

and

$$m_{ij} = m_{s-i,s-j} = \sum_{k=s-i}^{s-j-1} Q^s{}_k \quad \text{for } i > j.$$

From the vector α,

$$m_{ii} = \frac{2^s}{\binom{s}{i}}.$$

From these values we obtain

$$m_{i,i+1} + m_{i+1,i} = \frac{\sum_{k=0}^{i} \binom{s}{k}}{\binom{s-1}{i}} + \frac{\sum_{k=i+1}^{s} \binom{s}{k}}{\binom{s-1}{i}} = \frac{2^s}{\binom{s-1}{i}} \tag{1}$$

Let $n^{(i+1)}{}_{ii}$ be the mean number of times in state $\mathbf{V}_i$ before reaching state $\mathbf{V}_{i+1}$, the process is started in state $\mathbf{V}_i$. Then by § 6.2.7(b) we have

$$n^{(i+1)}{}_{ii} = \frac{m_{i,i+1} + m_{i+1,i}}{m_{ii}}$$
$$= \frac{2^s}{\binom{s-1}{i}} \frac{\binom{s}{i}}{2^s}$$
$$= \frac{s}{s-i}.$$

Assume now that s is even. From (1) we obtain

$$m_{0,s/2}+m_{s/2,0} = 2^s \sum_{i=0}^{s/2-1} \frac{1}{\binom{s-1}{i}}.$$

Also, since $m_{s/2,0}=m_{s/2,s}$, we have

$$m_{0,s} = m_{0,s/2}+m_{s/2,s} = 2^s \sum_{i=0}^{s/2-1} \frac{1}{\binom{s-1}{i}}. \tag{2}$$

From the point of view of physical interpretations the most interesting case is where there is a large number of molecules. For this we would like to find estimates for $m_{0,s/2}$ and $m_{s/2,0}$. The first represents the time taken to go from no molecules in the given region to an equal number in each. The second is the time taken to go from an equal number in each to 0 in region A. If the model is to have any similarity to actual physical situations we would expect that the time to equalize the molecules would be much shorter than the time to reach an extreme situation from equalization.

We consider first $m_{0,s/2}$. We estimate the time required to go from $\mathbf{V}_i$ to $\mathbf{V}_{i+1}$. We know that the mean number of times in $\mathbf{V}_i$ before reaching $\mathbf{V}_{i+1}$ is $s/(s-i)=1+i/(s-i)$. Each time it is in $\mathbf{V}_i$ (except for the last time) at least two steps are taken in going to $\mathbf{V}_{i-1}$ and back to $\mathbf{V}_i$. Hence the mean time to go from $\mathbf{V}_i$ to $\mathbf{V}_{i+1}$ is at least

$$1+2\frac{i}{s-i} = \frac{s+i}{s-i}.$$

Hence we may find a lower bound for $m_{0,s/2}$ by finding

$$\sum_{i=0}^{s/2-1} \frac{s+i}{s-i}.$$

This sum is greater than:

$$-2+\int_0^{s/2} \frac{s+x}{s-x}\,dx = s(2\log 2-{}^1\!/_2)-2.$$

Hence $s(2\log 2-{}^1\!/_2)-2$ is a lower bound for $m_{0,s/2}$. A better lower bound is obtained as follows. We assumed above that each time the process went from $\mathbf{V}_{i-1}$ to $\mathbf{V}_i$, it did so in one step. If instead we use our lower bound for the mean time to go from $\mathbf{V}_{i-1}$ to $\mathbf{V}_i$ we have a new lower bound for the time to go from $\mathbf{V}_i$ to $\mathbf{V}_{i+1}$. This is

$$1+\frac{i}{s-i}\left(1+\frac{s+i-1}{s-i+1}\right) = 1+\frac{2si}{(s-i)(s-i+1)}.$$

The sum of these values between $\mathbf{V}_0$ and $\mathbf{V}_{s/2-1}$ is greater than

$$-4+\int_0^{s/2}\left(1+\frac{2sx}{(s-x)(s-x+1)}\right)dx$$

$$= -4+s\left[{}^1/_2-2\log 2+s(s+1)\log\frac{s+2}{s+1}\right]$$

Using the first two terms in the Taylor series for $\log (s+2)/(s+1)$ we obtain the estimate

$$-5+s({}^5/_2-2\log 2)$$

for our improved lower bound.

We next obtain an upper bound for $m_{0,s/2}$. We have

$$m_{0,s/2} = \sum_{l=0}^{s/2-1}\frac{1}{\binom{s-1}{l}}\sum_{k=0}^{l}\binom{s}{k}$$

$$\sum_{k=0}^{l}\binom{s}{k} = \binom{s}{l}\left[1+\frac{l}{s-l+1}+\frac{l(l+1)}{(s-l+1)(s-l+2)}+\cdots+\frac{l!}{(s-l+1)\ldots s}\right]$$

$$\leqslant \binom{s}{l}\left[\sum_{j=0}^{\infty}\left(\frac{l}{s-l}\right)^j\right]$$

$$= \binom{s}{l}\frac{s-l}{s-2l}.$$

Hence,

$$m_{0,s/2} \leqslant \sum_{l=0}^{s/2-1}\frac{\binom{s}{l}}{\binom{s-1}{l}}\cdot\frac{s-l}{s-2l} = \sum_{l=0}^{s/2-1}\frac{s}{s-2l}$$

$$\leqslant \int_0^{(s-1)/2}\frac{s}{s-2x}\,dx+{}^1/_4s$$

$$= s({}^1/_4+{}^1/_2\log s).$$

This approximation can be improved by using a better estimate for the terms of the sum where l is large. Thus we have found that

$$-5+s({}^5/_2-2\log 2) < m_{0,s/2} < s[{}^1/_4+{}^1/_2\log s].$$

It would appear that these values are asymptotically of a greater order of magnitude than a constant times s, but less than a constant times $s\log s$. We see from these estimates that the process takes a very short time to go from 0 to $s/2$. For $s=100$ the minimum time it could take is 50 and the actual mean time required is about 140.

Consider next $m_{0,s}$. We know from (2) that

$$m_{0,s} = 2^s \sum_{i=0}^{s/2-1} \frac{1}{\binom{s-1}{i}}.$$

But

$$1+\frac{1}{s-1} \leqslant \sum_{i=0}^{s/2-1} \frac{1}{\binom{s-1}{i}} \leqslant 1+\frac{1}{s-1}+\frac{s/2-1}{\binom{s-1}{2}}.$$

Hence

$$2^s\left(1+\frac{1}{s-1}\right) \leqslant m_{0,s} \leqslant 2^s\left(1+\frac{2}{s-1}\right).$$

Thus

$$m_{0,s} \sim 2^s\left(1+\frac{A}{s}\right) \quad \text{where } 1 \leqslant A \leqslant 2.$$

Our estimates for $m_{0,s/2}$ show that this quantity is of a lower order of magnitude than $m_{0,s}$. Hence the above estimates serve also for $m_{s/2,s} = m_{s/2,0}$ and we have

$$m_{s/2,0} \sim 2^s\left(1+\frac{A}{s}\right) \quad \text{where } 1 \leqslant A \leqslant 2.$$

Thus, as predicted, the mean time to go from equalization to the extreme of 0 molecules is very much larger than to go from 0 to equalization. For $s = 100$ the first time is approximately 2^{100} or about 1000 billion billion billion while the second is only 140.

Other interesting quantities are $m_{0,0}$ and $m_{s/2,s/2}$. For these we have

$$m_{0,0} = 2^s$$

$$m_{s/2,s/2} = \frac{2^s}{\binom{s}{s/2}} \sim \sqrt{\frac{\pi}{2}s}.$$

For $s = 100$, $m_{50,50}$ is approximately 12.5.

We conclude this section with some remarks about the macroscopic chain and reversibility. It is sometimes argued that a process of the type we have considered has a "direction" because of the very great tendency to move toward equalization. It is true that if the process is started out of equilibrium—say, in state 0—then it will certainly move towards the center. However, if the process is observed after it has been going for a long time, then we may consider it as started in equilibrium and in this case the process will appear the same looked at in the reverse direction as in the forward. For example, if the process is then observed in $\mathbf{V}_x$ the probability that it moves to state $\mathbf{V}_{x-1}$ is the same as the probability that it came from $\mathbf{V}_{x-1}$. Another way to put this is the following. If a sequence of outcomes for a large number of steps is recorded and then handed to a physicist, he would be unable to tell whether he was given them in the order of increasing time or in the order of decreasing time.

§ 7.4 **Applications to genetics.** A problem that is frequently discussed in genetics is the following: Two animals are mated. From their offspring, two are selected by some method, and these are mated. Then the procedure is repeated. This type of problem may be treated as a Markov chain. As states we take the possible combinations of parents, and the transition probabilities are determined by the laws of genetics, from the assumptions concerning the way parents are selected.

The simplest such problem is obtained if we classify parents only according to the pair of genes they carry in one position in the chromosomes. We will discuss this case. Here we may further simplify the problem by assuming that the gene is either a or b, and hence that any individual animal must be of type aa or ab or bb. For example, if a dominates b, then aa is a pure dominant, ab is a hybrid, and bb is a pure recessive animal. Then a pair of parents must be of one of the following six types: (aa, aa), (bb, bb), (aa, ab), (bb, ab), (aa, bb), (ab, ab). This problem, for the simple case that the new parents are selected at random from the offspring, is treated in Feller and in FM.

We will discuss a class of problems of which the above problem is a special case. We will assume that one offspring is selected at random, and that this offspring selects a mate. In its selection it is k times as likely to pick a given animal unlike itself than a given animal like itself. Thus k measures how strongly "opposites attract each other." In this we take into account that in a simple dominance situation, aa and ab type animals are alike as far as appearances are concerned. The resulting transition matrix is

$$
P = \begin{array}{c} (aa, aa) \\ (bb, bb) \\ (aa, ab) \\ (bb, ab) \\ (aa, bb) \\ (ab, ab) \end{array}
\left(\begin{array}{cc|cccc}
1 & 0 & 0 & 0 & 0 & 0 \\
0 & 1 & 0 & 0 & 0 & 0 \\
\hline
1/4 & 0 & 1/2 & 0 & 0 & 1/4 \\
0 & \frac{1}{2(k+1)} & 0 & \frac{k}{(k+1)} & 0 & \frac{1}{2(k+1)} \\
0 & 0 & 0 & 0 & 0 & 1 \\
\frac{1}{4(k+3)} & \frac{1}{4(3k+1)} & \frac{1}{k+3} & \frac{2k(k+1)}{(k+3)(3k+1)} & \frac{k(k+1)}{(k+3)(3k+1)} & \frac{1}{k+3}
\end{array}\right).
$$

The first two states are absorbing. They correspond to having developed a pure strain: pure dominant in the first state, and pure recessive in the second. We will compute the fundamental matrix N, the vector τ, and the matrix B.

$$N = \frac{1}{(2k+1)(k+3)} \times \begin{pmatrix} 4(k^2+5k+2) & 2k(k+1)^2 & k(k+1) & (3k+1)(k+3) \\ 2(3k+1) & (4k^2+9k+3)(k+1) & k(k+1) & (3k+1)(k+3) \\ 4(3k+1) & 4k(k+1)^2 & (4k^2+9k+3) & 2(3k+1)(k+3) \\ 4(3k+1) & 4k(k+1)^2 & 2k(k+1) & 2(3k+1)(k+3) \end{pmatrix} \begin{matrix} (aa, ab) \\ (bb, ab) \\ (aa, bb) \\ (ab, ab) \end{matrix}$$

$$\tau = \frac{1}{(2k+1)(k+3)} \begin{pmatrix} 2k^3+12k^2+33k+11 \\ 4k^3+17k^2+29k+8 \\ 4k^3+18k^2+45k+13 \\ 4k^3+16k^2+38k+10 \end{pmatrix} \begin{matrix} (aa, ab) \\ (bb, ab) \\ (aa, bb) \\ (ab, ab) \end{matrix}$$

$$B = \frac{1}{4(2k+1)(k+3)} \begin{matrix} (aa, aa) & (bb, bb) \\ \end{matrix} \begin{pmatrix} 4k^2+23k+9 & 4k^2+5k+3 \\ 9k+3 & 8k^2+19k+9 \\ 18k+6 & 8k^2+10k+6 \\ 18k+6 & 8k^2+10k+6 \end{pmatrix} \begin{matrix} (aa, ab) \\ (bb, ab) \\ (aa, bb) \\ (ab, ab) \end{matrix}.$$

Since we know that we will eventually end up with a pure strain, the two most interesting questions concern the number of generations needed to reach a pure strain and the probability of getting pure dominants or pure recessives. In particular, we will be interested in the effect that k has on these quantities.

A large k has the effect of producing more mixed matings. Hence we would expect a large k to slow down the process. Indeed, every entry in τ is monotone increasing in k. Some typical values of this vector are

$$\underset{k=0}{\begin{pmatrix} 3^2/_3 \\ 2^2/_3 \\ 4^1/_3 \\ 3^1/_3 \end{pmatrix}} \quad \underset{k=1}{\begin{pmatrix} 4^5/_6 \\ 4^5/_6 \\ 6^2/_3 \\ 5^2/_3 \end{pmatrix}} \quad \underset{k=2}{\begin{pmatrix} 5.64 \\ 6.64 \\ 8.28 \\ 7.28 \end{pmatrix}} \quad \underset{k=7}{\begin{pmatrix} 10.11 \\ 16.11 \\ 17.21 \\ 16.21 \end{pmatrix}}.$$

We see that increasing k will slow down the process considerably, especially if we start in one of the last three states. The fact that the time to absorption from (aa, bb) is always one more than from

(ab, ab) is due to the fact that from the former state we always go to the latter in one step (a pure dominant and a pure recessive parent must have hybrid offspring).

The effect of k on the probability of absorption is not so clear. One would guess that large k favors the recessive strain, since then both dominants and hybrids would tend to select recessive mates. Indeed, for $k>1$ the probability of absorption in (bb, bb) increases with k. But for $k<1$ a surprising situation develops. As k is decreased from 1, the probability of absorption in (aa, aa) increases, until it reaches a maximum at $k={}^1/_3$, and then decreases back to the same value as at $k=1$. Thus $k=0$ and $k=1$ yield the same absorption probabilities. This means that if in mating like always selects like, the probability of absorption is the same as for random mating (though of course the time to absorption is much less than for random mating). Some typical values for the probability of absorption in (aa, aa) are

$$\begin{pmatrix} .75 \\ .25 \\ .50 \\ .50 \end{pmatrix} \quad \begin{pmatrix} .77 \\ .27 \\ .54 \\ .54 \end{pmatrix} \quad \begin{pmatrix} .75 \\ .25 \\ .50 \\ .50 \end{pmatrix} \quad \begin{pmatrix} .71 \\ .21 \\ .42 \\ .42 \end{pmatrix} \quad \begin{pmatrix} .61 \\ .11 \\ .22 \\ .22 \end{pmatrix}.$$
$$k=0 \qquad k={}^1/_3 \qquad k=1 \qquad k=2 \qquad k=7$$

The fact that the last two entries are always the same is due to the already observed fact that the process always goes directly from (aa, bb) to (ab, ab). The probability from (bb, ab) is half this much; since when the process leaves (bb, ab), it is equally likely to go to (ab, ab) or to be absorbed in (bb, bb). Similarly, the value at (aa, ab) is half the (ab, ab) value, plus ${}^1/_2$. It is interesting to note that for random mating the probability of absorption in (aa, aa) is proportional to the number of a genes present at the start.

Let us collect the quantities for the special case of random mating, $k=1$.

$$P = \left(\begin{array}{cc|cccc} 1 & 0 & 0 & 0 & 0 & 0 \\ 0 & 1 & 0 & 0 & 0 & 0 \\ \hline {}^1/_4 & 0 & {}^1/_2 & 0 & 0 & {}^1/_4 \\ 0 & {}^1/_4 & 0 & {}^1/_2 & 0 & {}^1/_4 \\ 0 & 0 & 0 & 0 & 0 & 1 \\ {}^1/_{16} & {}^1/_{16} & {}^1/_4 & {}^1/_4 & {}^1/_8 & {}^1/_4 \end{array}\right) \begin{array}{l} (aa, aa) \\ (bb, bb) \\ \\ (aa, ab) \\ (bb, ab) \\ (aa, bb) \\ (ab, ab) \end{array}$$

$$N = \begin{pmatrix} 8/3 & 2/3 & 1/6 & 4/3 \\ 2/3 & 8/3 & 1/6 & 4/3 \\ 4/3 & 4/3 & 4/3 & 8/3 \\ 4/3 & 4/3 & 1/3 & 8/3 \end{pmatrix}$$

$$\tau = \begin{pmatrix} 4^5/_6 \\ 4^5/_6 \\ 6^2/_3 \\ 5^2/_3 \end{pmatrix} \qquad B = \begin{pmatrix} 3/4 & 1/4 \\ 1/4 & 3/4 \\ 1/2 & 1/2 \\ 1/2 & 1/2 \end{pmatrix}.$$

We also compute

$$H = \begin{pmatrix} 5/8 & 1/4 & 1/8 & 1/2 \\ 1/4 & 5/8 & 1/8 & 1/2 \\ 1/2 & 1/2 & 1/4 & 1 \\ 1/2 & 1/2 & 1/4 & 5/8 \end{pmatrix} \qquad \tau_2 = 1/36 \begin{pmatrix} 769 \\ 769 \\ 816 \\ 816 \end{pmatrix}.$$

The standard deviation in τ is around 4.7 for every entry. This is of the same order of magnitude as the entries of τ, hence we may expect very large fluctuations.

We observe from P that the process satisfies the condition for lumpability if we combine the first two states and combine the next two states. This partition has a simple interpretation. The state (bb, bb) results from (aa, aa) by interchanging a and b, and (bb, ab) results from (aa, ab) similarly. On the other hand (aa, bb) and (ab, ab) are unchanged. Hence the partition represents the process if we do not care which gene is the dominant gene. The first state of the new process, which we may denote by (aa, aa), represents any pair of like pure parents. The second state, to be denoted by (aa, ab), represents one pure and one hybrid parent. The remaining two states represent the combination of unlike pure parents, and of two hybrid parents, respectively. The lumped transition matrix is

$$\hat{P} = \begin{pmatrix} 1 & 0 & 0 & 0 \\ 1/4 & 1/2 & 0 & 1/4 \\ 0 & 0 & 0 & 1 \\ 1/8 & 1/2 & 1/8 & 1/4 \end{pmatrix} \begin{matrix} (aa, aa) \\ (aa, ab) \\ (aa, bb) \\ (ab, ab) \end{matrix}.$$

To obtain N for the lumped process, we add the first two columns, which correspond to lumped transient states:

$$\begin{pmatrix} 10/3 & 1/6 & 4/3 \\ 10/3 & 1/6 & 4/3 \\ 8/3 & 4/3 & 8/3 \\ 8/3 & 1/3 & 8/3 \end{pmatrix}.$$

We then observe that the first two rows are identical, as must be the case for lumpability. Hence we have

$$\hat{N} = \begin{pmatrix} 10/3 & 1/6 & 4/3 \\ 8/3 & 4/3 & 8/3 \\ 8/3 & 1/3 & 8/3 \end{pmatrix} \quad \begin{matrix} (aa, ab) \\ (aa, bb) \\ (ab, ab) \end{matrix}$$

and

$$\hat{\tau} = \hat{N}\xi = \begin{pmatrix} 4\,5/6 \\ 6\,2/3 \\ 5\,2/3 \end{pmatrix}.$$

The vector $\hat{\tau}$ is also obtainable directly from τ. The latter has identical entries for the two states to be lumped, as must be the case, and contraction yields $\hat{\tau}$. For many purposes the lumped process yields sufficient information. (Of course, it does not yield any interesting information as far as the absorption probabilities are concerned, since there is only one absorbing state after lumping.) For example, τ is completely determined by $\hat{\tau}$.

$$\hat{H} = \begin{pmatrix} 7/10 & 1/8 & 1/2 \\ 8/10 & 1/4 & 1 \\ 8/10 & 1/4 & 5/8 \end{pmatrix} \qquad \hat{\tau}_2 = \begin{pmatrix} 21\,13/36 \\ 22\,2/3 \\ 22\,2/3 \end{pmatrix}.$$

The last two columns of $\hat{H}$, corresponding to single-state cells, are obtainable directly from H. But the first column is new. $\hat{\tau}_2$ is directly obtainable from τ_2.

A generalization of this lumped process is discussed in Kempthorne.† We still restrict ourselves to a single position in the chromosomes, but we no longer assume that only two types of genes can occur in this position. There may be any number of different kinds of genes. We consider the process in the lumped form: We care about the number of different genes present and about their combination, but we do not distinguish states that differ only in having the genes permuted. Thus

† O. Kempthorne, *An Introduction to Genetic Statistics*, New York, John Wiley & Son, Inc., 1950.

we have the following seven states:

s_1: (aa, aa)
s_2: (aa, ab)
s_3: (aa, bb)
s_4: (ab, ab)
s_5: (aa, bc)
s_6: (ab, ac)
s_7: (ab, cd)

The transition matrix is:

$$P = \left(\begin{array}{c|ccc|cc|c} 1 & 0 & 0 & 0 & 0 & 0 & 0 \\ \hline 1/4 & 1/2 & 0 & 1/4 & 0 & 0 & 0 \\ 0 & 0 & 0 & 1 & 0 & 0 & 0 \\ 1/8 & 1/2 & 1/8 & 1/4 & 0 & 0 & 0 \\ \hline 0 & 0 & 0 & 1/2 & 0 & 1/2 & 0 \\ 1/16 & 1/4 & 0 & 3/16 & 1/8 & 3/8 & 0 \\ \hline 0 & 0 & 0 & 1/4 & 0 & 1/2 & 1/4 \end{array}\right) \begin{array}{c} s_1 \\ s_2 \\ s_3 \\ s_4 \\ s_5 \\ s_6 \\ s_7 \end{array}$$

We have indicated the equivalence classes in the transition matrix. The equivalence classes are determined by the number of different genes present in the parents. Clearly, this number either stays the same or it decreases, hence the process can move from an equivalence class with a given number of genes only to one with fewer genes, i.e. from the bottom up in P. The single state with only one type of gene present is absorbing. The four-state lumped process we considered above corresponds to the two top equivalence classes in the present chain. Since numbers concerning these classes are not affected by equivalence classes lower down, all quantities we are about to compute will agree in their upper left corner with those previously found.

$$N = \left(\begin{array}{ccc|cc|c} 10/3 & 1/6 & 4/3 & 0 & 0 & 0 \\ 8/3 & 4/3 & 8/3 & 0 & 0 & 0 \\ 8/3 & 1/3 & 8/3 & 0 & 0 & 0 \\ \hline 8/3 & 5/18 & 20/9 & 10/9 & 8/9 & 0 \\ 8/3 & 2/9 & 16/9 & 2/9 & 16/9 & 0 \\ \hline 8/3 & 7/27 & 56/27 & 4/27 & 32/27 & 4/3 \end{array}\right) \begin{array}{c} s_2 \\ s_3 \\ s_4 \\ s_5 \\ s_6 \\ s_7 \end{array}$$

$$\tau = \begin{pmatrix} 4^5/_6 \\ 6^2/_3 \\ 5^2/_3 \\ 7^1/_6 \\ 6^2/_3 \\ 7^2/_3 \end{pmatrix} \begin{matrix} s_2 \\ s_3 \\ s_4 \\ s_5 \\ s_6 \\ s_7 \end{matrix} \qquad \tau_2 = \begin{pmatrix} 21^{13}/_{36} \\ 22^2/_3 \\ 22^2/_3 \\ 23^{11}/_{12} \\ 24^2/_3 \\ 24^2/_3 \end{pmatrix}$$

$$H = \begin{matrix} s_2 & s_3 & s_4 & s_5 & s_6 & s_7 \end{matrix}$$
$$H = \begin{pmatrix} 7/10 & 1/8 & 1/2 & 0 & 0 & 0 \\ 8/10 & 1/4 & 1 & 0 & 0 & 0 \\ 8/10 & 1/4 & 5/8 & 0 & 0 & 0 \\ 8/10 & 5/24 & 5/6 & 1/10 & 1/2 & 0 \\ 8/10 & 1/6 & 2/3 & 1/5 & 7/16 & 0 \\ 8/10 & 7/36 & 7/9 & 2/15 & 2/3 & 1/4 \end{pmatrix}.$$

We note that s_2, s_4, and s_6 are the "likely states," in the sense that if the process starts low enough for it to be possible to reach the state there is always a fairly good chance of reaching the state. The other states are quite unlikely no matter how the process is started. It is quite surprising that starting in s_7—that is, starting with four different genes—we expect to reach a pure strain in $7^2/_3$ generations. But it must be remembered that the standard deviation of this quantity is 4.97, and hence much longer processes are not too unlikely.

§ **7.5 Learning theory.** This section will be devoted to the study of a mathematical model for certain kinds of learning, due to W. K. Estes. We will discuss only some relatively simple special cases, but the techniques here used are applicable to more general situations.

In a typical experiment the subject is placed in front of a pair of lights, and he is asked to guess whether the light on his left or the light on his right will be turned on next. Thus, he has two possible responses. We denote by $\mathbf{A}_0$ the guess "left," and by $\mathbf{A}_1$ the guess "right." Then the experimenter turns on one of the lights. Let $\mathbf{E}_0$ mean turning on the left light, and $\mathbf{E}_1$ the right light. The procedure is repeated a large number of times, and a record is kept of the sequence of both $\mathbf{A}_t$ and $\mathbf{E}_t$. The purpose of the theory is to predict, for given behavior of the experimenter, how the subject's guesses will change in the long run. A variety of experiments has shown that the model is in good agreement with the facts.

In a large class of interesting experiments the experimenter will choose his actions with fixed probabilities, depending only on the action of the subject. These probabilities may be given in Figure 7-8.

$$\begin{array}{c} \\ \mathbf{A}_0 \\ \mathbf{A}_1 \end{array}\begin{array}{c} \begin{array}{cc} \mathbf{E}_0 & \mathbf{E}_1 \end{array} \\ \begin{pmatrix} 1-v & v \\ w & 1-w \end{pmatrix} \end{array}$$

FIGURE 7-8

That is, a guess "left" ($\mathbf{A}_0$) is *reinforced* by turning on the left light ($\mathbf{E}_0$) with probability $1-v$; otherwise the right light is turned on. And a guess "right" is reinforced with probability $1-w$. Here v and w are numbers between 0 and 1, which are kept fixed for the duration of the experiment.

While v and w are usually positive numbers, the cases in which they are not lead to interesting experiments. For example, if $v=0$ and $w>0$, then $\mathbf{A}_0$ is always reinforced, but $\mathbf{A}_1$ is only occasionally reinforced. The case $v>0$ and $w=0$ is similar. If $v=w=0$, then every action of the subject is reinforced.

Another class of interesting special cases is where $v+w=1$. Here $1-v=w$ and $v=1-w$, and hence the probability of E_0 or of E_1 is independent of the action of the subject.

The model assumes that the subject has a certain unknown number s of *stimulus elements*. Each stimulus element is at each stage of the experiment connected to one of the two possible responses $\mathbf{A}_i$, the original connections not being known. It is then assumed that the following takes place at each stage of the experiment:

(1) The subject "samples" a subset of the stimulus elements, by means of an independent trials process, in which any one stimulus element is sampled with probability t or not sampled with probability $1-t$.

(2) If in the sampled set there are k stimulus elements connected to $\mathbf{A}_0$ and l to $\mathbf{A}_1$, then the subject performs $\mathbf{A}_1$ with probability $l/(k+l)$. Some convention is necessary if no stimulus element is sampled; we will assume that the probability of $\mathbf{A}_1$ is the same in this case as if *all* stimulus elements had been sampled.

(3) If the experimenter performs $\mathbf{E}_0$, then any stimulus element that was previously connected to $\mathbf{A}_1$ and which was just sampled by the subject is reconnected to $\mathbf{A}_0$. Similarly, after $\mathbf{E}_1$, all the sampled stimulus elements are connected to $\mathbf{A}_1$.

We can represent the model as an $(s+1)$ state Markov chain, in which state $\mathbf{s}_i$ occurs when exactly i stimulus elements are connected to $\mathbf{A}_1$, and $i = 0, 1, \ldots, s$. All the interesting quantities depend only on how many stimulus elements are connected each way, and hence these quantities are functions on the chain. For example, the probability of action $\mathbf{A}_1$ from state $\mathbf{s}_i$ is obtained as follows:

$$\mathbf{Pr}_i[\mathbf{A}_1] = \sum_{k=0}^{s-i} \sum_{\substack{l=0 \\ k,\, l \text{ not both } 0}}^{i} \binom{s-i}{k}\binom{i}{l} t^m(1-t)^{s-m}\frac{l}{m} + (1-t)^s\frac{i}{s},$$

where k is the number of stimulus elements sampled from those connected to $\mathbf{A}_0$, l from those connected to $\mathbf{A}_1$, $m = k + l$, and the last term arises from the assumption concerning the case where no stimulus element is sampled. We rewrite this as a sum on m, and then use a binomial identity:

$$\begin{aligned}
\sum_{m=1}^{s} t^m(1-t)^{s-m} \sum_{k+l=m} \binom{s-1}{k}\binom{i}{l}\frac{l}{m} + (1-t)^s\frac{i}{s} \\
&= \sum_{m=1}^{s} t^m(1-t)^{s-m}\binom{s}{m}\frac{i}{s} + (1-t)^s\frac{i}{s} \\
&= \frac{i}{s}\sum_{m=0}^{s}\binom{s}{m}t^m(1-t)^{s-m} \\
&= \frac{i}{s}.
\end{aligned}$$

It will be convenient to let the column vector $\gamma = \left\{\frac{i}{s}\right\}$ represent these probabilities.

Let us next construct the transition matrix P. For this we must take into account all four possibilities in Figure 7-8. The combination of $\mathbf{A}_1$ and $\mathbf{E}_0$, with a transition from $\mathbf{s}_i$ *down* to $\mathbf{s}_j$ has probabilities wX, where

$$x_{ij} = \begin{cases} \displaystyle\sum_{k=0}^{s-i}\binom{s-i}{k}\binom{i}{i-j}t^{i-j+k}(1-t)^{s-i+j-k}\cdot\frac{i-j}{i-j+k} & \text{if } j < i \\ 0 & \text{if } j \geqslant i \end{cases}$$

The downward transition $\mathbf{s}_i$ to $\mathbf{s}_j$, by means of the combination $\mathbf{A}_0$ and $\mathbf{E}_0$, has probabilities $(1-v)(Y-X)$, where

$$y_{ij} = \begin{cases} \displaystyle\binom{i}{i-j}t^{i-j}(1-t)^j & \text{if } j \leqslant i \\ 0 & \text{if } j > i \end{cases}$$

If we let $x^*_{ij} = x_{s-i,\, s-j}$, and $y^*_{ij} = y_{s-i,\, s-j}$, then vX^* and $(1-w)(Y^*-X^*)$ represent the probabilities of upwards transition. Thus

$$P = wX + vX^* + (1-v)(Y-X) + (1-w)(Y^*-X^*) + (v+w-1)(1-t)^s I,$$

or

$$P = v(X+X^*-Y) + w(X+X^*-Y^*) + (Y+Y^*) - (X+X^*) + (v+w-1)(1-t)^s I. \tag{1}$$

The final term represents the cases where no stimulus element is sampled, which were not included in the two previous terms.

We wish to compute $P\gamma$. First we find $Y\gamma$.

$$\begin{aligned}\sum_{k=0}^{s} y_{ik}\frac{k}{s} &= \sum_{k=0}^{i}\binom{i}{i-k}t^{i-k}(1-t)^k\frac{k}{s}\\ &= \frac{1}{s}\sum_{k=1}^{i} k\binom{i}{k}t^{i-k}(1-t)^k\\ &= \frac{i}{s}\sum_{k=1}^{i}\binom{i-1}{k-1}t^{i-k}(1-t)^k\\ &= \frac{i}{s}\sum_{l=0}^{i-1}\binom{i-1}{l}t^{i-1-l}(1-t)^{l+1}\\ &= \frac{i}{s}(1-t).\end{aligned}$$

Hence

$$Y\gamma = (1-t)\gamma. \tag{2}$$

Quite similarly,

$$Y^*\gamma = t\xi + (1-t)\gamma. \tag{3}$$

And by a somewhat longer argument we find that

$$(X+X^*)\gamma = t\xi + (1-2t-(1-t)^s)\gamma. \tag{4}$$

If we write $P\gamma$ by means of (1), and make use of (2), (3), and (4), we obtain

$$\begin{aligned}P\gamma = {} & v[t\xi + (1-2t-(1-t)^s)\gamma - (1-t)\gamma]\\ & + w[t\xi + (1-2t-(1-t)^s)\gamma - t\xi - (1-t)\gamma]\\ & + t\xi + 2(1-t)\gamma - t\xi - (1-2t-(1-t)^s)\gamma\\ & + (v+w-1)(1-t)^s\gamma.\end{aligned}$$

This simplifies to

$$P\gamma = vt\xi + [1-(v+w)t]\gamma. \tag{5}$$

Let us introduce the vector $\delta = \gamma - \left(\frac{v}{v+w}\right)\xi$ for the cases where v

and w are not both 0. It will be shown later that δ has an important interpretation in the model.

$$P\delta = P\gamma - \left(\frac{v}{v+w}\right)\xi = [1-(v+w)t]\gamma - \left(\frac{v}{v+w}\right)[1-(v+w)t]\xi$$

or

$$P\delta = [1-(v+w)t]\delta \tag{6}$$

and

$$P^n\delta = [1-(v+w)t]^n\delta. \tag{7}$$

The values of v and w determine the nature of the Markov chain. If $0 < v \leqslant 1$ and $0 < w \leqslant 1$, then $P > 0$ and hence the chain is regular. If $v = 0$, then $i = 0$ is an absorbing state; and if $w = 0$, then $i = s$ is an absorbing state. Hence if either v or w is 0, we have an absorbing chain. If $v = w = 0$, then we have two absorbing states.

Let us find the probability of an $\mathbf{A}_1$ response after n steps. This will be given by the vector $P^n\gamma$, the probability depending on the starting state. If $v = w = 0$, then $P\gamma = \gamma$ from (5), hence $P^n\gamma = \gamma$. The probability of an $\mathbf{A}_1$ response is unchanged. In the other cases, (7) provides the answer

$$P^n\gamma = [v/(v+w)]\xi + [1-(v+w)t]^n\delta.$$

The first term (as will be seen below) is the limiting probability for an $\mathbf{A}_1$ response, and the second is the deviation due to the initial position.

Let us first discuss the regular case. Here $v > 0$ and $w > 0$. Since $0 < t < 1$, we have $|1-(v+w)t| < 1$. From (7),

$$A\delta = 0$$

$$\alpha\delta = \alpha\left(\gamma - \left(\frac{v}{v+w}\right)\xi\right) = 0$$

and thus

$$\alpha\gamma = \frac{v}{v+w}. \tag{8}$$

This proves that the limiting probability of an $\mathbf{A}_1$ response is $v/(v+w)$. Applying this to Figure 7-8 we find that the limiting probability of an $\mathbf{E}_1$ action by the experimenter is

$$\frac{w}{v+w}\cdot v + \frac{v}{v+w}\cdot(1-w) = \frac{v}{v+w}.$$

Thus in equilibrium the probabilities for the experimenter and the subject are in agreement.

It is interesting to note that the subject does *not* maximize the number of correct guesses. Instead, he brings about an equilibrium in which he is guessing "right" with the same frequency with which "right" comes up.

Since $v/(v+w)$ is the mean number of $\mathbf{A}_1$ responses per trial in equilibrium, and since i/s is the mean number when in state $\mathbf{s}_i$, the vector $\delta = \left\{\frac{i}{s} - \left(\frac{v}{v+w}\right)\right\}$ gives the deviation between the mean number of $\mathbf{A}_1$ responses in a given state and in equilibrium.

From (6), (7),

$$(I-P+A)\delta = (v+w)t\delta$$

$$Z\delta = \frac{1}{(v+w)t}\delta$$

$$(Z-A)\delta = \frac{1}{(v+w)t}\delta$$

$$(Z-A)\gamma = \frac{1}{(v+w)t}\delta. \tag{9}$$

Thus the total deviation from equilibrium (for the number of $\mathbf{A}_1$ responses) is proportional to the deviation vector δ. Hence the total deviation may be large because i/s for the starting state may have been far from the equilibrium $v/(v+w)$, or because t is small.

To obtain the limiting variance for the number of $\mathbf{A}_1$ responses, we must use § 4.6, since we have a function on the chain which takes on the value 1 in $\mathbf{s}_i$ with probability $\mathbf{f}_i = i/s$. While we have no general formula for this limiting variance, in any concrete example it is easy to compute it from § 4.6.

Let us next consider the case of an absorbing chain with one absorbing state $i=0$, that is, $v=0$ and $w>0$. (The case $v>0$, $w=0$ is similar.) In this case $\delta = \gamma$. Let $\bar{\gamma}$ be gotten from γ by deleting its first component (which is 0). From (6),

$$P\gamma = P\delta = (1-wt)\delta = (1-wt)\gamma$$

$$Q\bar{\gamma} = (1-wt)\bar{\gamma}$$

$$(I-Q)\bar{\gamma} = (wt)\bar{\gamma}$$

$$N\bar{\gamma} = \frac{1}{wt}\bar{\gamma}. \tag{10}$$

This case is one in which the subject is being conditioned to give $\mathbf{A}_0$ responses. If he gives an $\mathbf{A}_0$ response, it is always reinforced. But an $\mathbf{A}_1$ response is also reinforced occasionally, with probability $1-w$. $N\bar{\gamma}$

gives the mean of the total number of $\mathbf{A}_1$ (or "wrong") responses. For starting state $\mathbf{s}_i$ this is $1/wt \cdot i/s$. Thus there may be a large number of errors for one of three reasons: The fraction i/s of stimulus elements that need reconditioning was high at the start; or the learning parameter t is low; or w is low, that is, an $\mathbf{A}_1$ response is frequently reinforced.

It is sometimes reasonable to assume that the stimulus elements were originally connected at random. This gives us an initial probability vector $\pi = \left\{\frac{1}{2^s}\binom{s}{i}\right\}$. Then the mean of the total number of "wrong" responses is

$$\pi \cdot \frac{1}{wt}\gamma = \left(\frac{1}{wt}\right)\pi\gamma = \frac{1}{2wt}.$$

We could expect to average this number of $\mathbf{A}_1$ responses in a large homogeneous population. This is one simple way of estimating t.

Finally we discuss the case $v = w = 0$, where every action of the subject is reinforced. Then both $i = 0$ and $i = s$ are absorbing states, and the most interesting question is whether the subject ends up conditioned to $\mathbf{A}_0$ or to $\mathbf{A}_1$ responses.

From (5) we see that in this case $P\gamma = \gamma$. But γ has a 0 component for $\mathbf{s}_0$, and 1 for $\mathbf{s}_s$, hence (see Theorem 3.3.9) γ gives the probabilities for absorption in $\mathbf{s}_s$. Thus the probability of being conditioned entirely to $\mathbf{A}_1$ responses is equal to the fraction of stimulus elements originally connected to $\mathbf{A}_1$.

To obtain more detailed information about the model we will have to make a simplifying assumption. Since in most applications the learning parameter t is quite small, we shall assume that terms of higher order in t are negligible. This may also be interpreted psychologically. If we drop terms in powers of t higher than the first, we assume that the sampling of more than one stimulus element at a time is quite unlikely. Under this assumption P is considerably simplified:

$$\begin{aligned} p_{i,i+1} &= (s-i)tv \\ p_{i,i-1} &= itw \\ p_{ii} &= 1 - t(iw + (s-i)v) \\ p_{ij} &= 0 \quad \text{otherwise.} \end{aligned}$$

Let us consider first the regular case under this simplifying assumption. The most important quantity lacking in our previous treatment was the vector of limiting probabilities α. We shall show that under our present assumption

$$a_i = \binom{s}{i} v^i w^{s-i}/(v+w)^s.$$

Let us compute αP.

$$\sum_{k=0}^{s} a_k p_{kj} = [a_{j-1}p_{j-1,j} + a_j p_{jj} + a_{j+1}p_{j+1,j}]$$

$$= \left[\binom{s}{j-1} v^{j-1}w^{s-j+1}(s-j+1)tv + \binom{s}{j} v^j w^{s-j}[1-t(jw+(s-j)v)]\right.$$

$$\left. + \binom{s}{j+1} v^{j+1}w^{s-j-1}(j+1)tw\right] \Big/ (v+w)^s$$

$$= a_j + \frac{tv^j w^{s-j}}{(v+w)^s}\left[\binom{s}{j-1}(s-j+1)w - \binom{s}{j}(jw+(s-j)v)\right.$$

$$\left. + \binom{s}{j+1}(j+1)v\right].$$

We then find that the two v-terms cancel each other and the two w-terms also cancel, and hence the expression in brackets is 0. Thus the right side reduces to a_j, and hence $\alpha = \{a_i\}$ is the fixed vector of P. Thus we know that if t is small, the limiting probabilities are very nearly given by this α. Indeed α is the limit of the limiting probabilities as $t \to 0$.

It is interesting to note that for $w = v = 1/st$ the process we obtain is the same as that obtained for the macroscopic process in the Ehrenfest model.

Next let us consider an absorbing case, say $w > 0$ and $v = 0$, that is, where the subject is being conditioned to response $\mathbf{A}_0$. Then

$$p_{i,i-1} = itw$$
$$p_{ii} = 1 - itw$$

and all other entries are 0. The matrix $I - Q = \{c_{ij}\}$ where $c_{ii} = itw$, and $c_{i,i-1} = -itw$, and all other entries are 0. We will show that

$$n_{ij} = 1/jtw \quad \text{if } j \leqslant i$$
$$= 0 \text{ otherwise.}$$

Let us compute $N(I-Q)$.

$$\sum_{k=1}^{s} n_{ik}c_{kj} = n_{ij}(jtw) + n_{i,j+1}[-(j+1)tw]$$

if $i < j$: $= 0 + 0 = 0$

if $i = j$: $= (1/jtw)jtw + 0 = 1$

if $i > j$: $= (1/jtw)jtw + (1/(j+1)tw)(-(j+1)tw) = 1 - 1$
$= 0,$

which verifies that N is the desired inverse. It is interesting to note that N depends on i, j, t, but not on s. We then obtain

$$t_i = (1/tw) \sum_{j=1}^{i} (1/j).$$

Thus the time of conditioning is inversely proportional to t and to w, and depends on the number of stimulus elements that need to be conditioned from $\mathbf{A}_1$ to $\mathbf{A}_0$. The time will be large if t is small. It will be large if w is small—that is, $\mathbf{A}_1$ is frequently reinforced. But since the series $1/j$ diverges, the time may also be large because many stimuli elements were originally conditioned to $\mathbf{A}_1$.

Finally we shall consider the case of two absorbing states, $w=v=0$. Here the first-order terms in t drop out, and hence we will have to carry out our computation in terms of t^2.

$$\begin{aligned} p_{i,i-1} &= p_{i,i+1} = (1/2)i(s-i)t^2 \\ p_{ii} &= 1-i(s-i)t^2. \end{aligned}$$

A computation quite similar to the one above will verify that

$$\begin{aligned} n_{ij} &= (2/t^2 s)(i/j) & \text{if } i \leqslant j. \\ &= (2/t^2 s)(s-i)/(s-j) & \text{if } i \geqslant j. \end{aligned}$$

and hence

$$t_i = (2/t^2)\left[\left(1-\frac{i}{s}\right)\sum_{j=s-i}^{s-1}\frac{1}{j}+\frac{i}{s}\sum_{j=i+1}^{s-1}\frac{1}{j}\right].$$

Again the sum may be large because t is small or because there are many terms (s is large). But in this case t_i is inversely proportional to t^2, and hence we expect a much longer time for conditioning.

There is one special case in which more precise information is available without the above simplifying assumption. This is the case where $v+w=1$, or $w=1-v$. Here the action of the experimenter is independent of what the subject does. We found an exact solution for α in this case, in terms of simple recursion equations.†

It was shown that the limiting probabilities may also be obtained from the following auxiliary process: We start with s stimuli elements completely unconditioned. We select a subset of these, picking each stimulus element with probability t. Then by a random device we assign these to $\mathbf{A}_0$ with probability w or to $\mathbf{A}_1$ with probability $1-w$. We then apply the same procedure to the remaining stimuli elements, till all are assigned. Then the limiting probability $\mathbf{a}_i$ for the original process is simply the probability of assigning i stimuli elements to $\mathbf{A}_1$

† Cf. J. G. Kemeny and J. L. Snell, "Markov Processes in Learning Theory," *Psychometrika*, **22** (No. 3):221–230, 1957.

in our auxiliary process. These probabilities are easily obtained for any s of reasonable size.

§ 7.6 Applications to mobility theory. In this section we shall consider the application of Markov chain ideas to a problem in sociology, the problem of intergenerational occupational mobility. The results of this section were prepared jointly with J. Berger. The problem may be stated as follows. A partition $\mathbf{A} = \{\mathbf{A}_1, \mathbf{A}_2, \ldots, \mathbf{A}_r\}$ is made of the set of all occupations. The cells are called the *occupational classes.* These are usually ordered with respect to some socially relevant criterion, for example, prestige of occupation. The question is then asked, to what extent does the occupational class of the father, grandfather, etc., affect the occupational class of the son? In any such study a matrix is constructed which represents for each class the fraction of the sons that would be expected to go into each of the occupations.

We shall take as our basic example a matrix constructed from data collected by Glass and Hall from England and Wales for 1949.† Following Prais,‡ we classify the occupations as upper, middle, and lower. The estimated matrix is

$$P = \begin{matrix} & \text{UPPER} & \text{MIDDLE} & \text{LOWER} \\ \text{UPPER} \\ \text{MIDDLE} \\ \text{LOWER} \end{matrix} \begin{pmatrix} .448 & .484 & .068 \\ .054 & .699 & .247 \\ .011 & .503 & .486 \end{pmatrix}.$$

We see, for example, that of the upper class, 44.8 percent of the sons went to the upper class, 48.4 percent to the middle, and 6.8 percent to the lower.

There are two ways to make use of P. One is to consider at each state the total population, and predict the fraction of the population which will be in each of the occupational classes. We shall call this the "collective process".

A second way is to study a single family history. From this point of view we consider this history as the outcomes of a Markov chain with transition matrix P. We shall call this the "individual process." We assume every family has exactly one son.

We proceed now to discuss the relation of the basic concepts of Markov chain theory to these two processes.

We begin with the assumptions for a Markov chain. The basic

† D. V. Glass and J. R. Hall, "Social Mobility in Great Britain: A Study of Intergeneration Changes in Status," in D. V. Glass (Ed.), *Social Mobility in Great Britain*, London, Routledge & Kegan Paul, 1954.

‡ S. J. Prais, "Measuring Social Mobility," *Journal of the Royal Statistical Society*, **118**:56–66, 1955.

assumption is that the knowledge of the past beyond the last outcome does not influence our predictions. In the individual process this would mean, for example, that the knowledge of the occupation of the grandfather would not affect our predictions for the son.

We also assume that the same P serves for every generation. This is clearly not completely realistic. However, there is still a great deal of interest in studying what would happen if the present P were to continue to be appropriate.

It is also assumed that changes in distributions in occupational classes from one generation to the next are to be accounted for only by the process described by P; that is to say, for purposes of this analysis we ignore the effect of differential reproduction and migration rates, as these may be related to occupations of the system.

The classification of states has obvious interpretations in mobility. An ergodic set is a set of occupations from which it is impossible to leave. In most industrialized societies we would expect only one ergodic set. However, if we take as state the pair, occupation and race, then discrimination against a certain race may cause the resulting chain to have more than one ergodic set. When studying occupations, a son can have the same occupation as his father. Thus we would not expect cyclic chains. An absorbing state would mean that for a given occupation the son must follow his father's footsteps. Again, in industrialized societies occupations do not usually have this property. We shall therefore assume that our basic chain is regular.

Let us next see the interpretation of the powers of P. For the individual process the ij-th entry of P^n will give the probability that, after n generations the family will be in the j-th occupation class if it started in the i-th. For the collective process, $p^{(n)}_{ij}$ represents the fraction of the descendants of people in the i-th occupational class that will be in the j-th occupational class after n generations. If we start with fractions $\pi = (p_1, p_2, p_3)$ in each of the classes, then after n generations there will be fractions given by πP^n. In our example, assume that there are at present 20 percent in the upper class, 70 percent in the middle class, and 10 percent in the lower class. Then after one generation the percentages are 12.9, 63.6, and 23.5. We obtain these by

$$(.200 \quad .700 \quad .100)\begin{pmatrix} .448 & .484 & .068 \\ .054 & .699 & .247 \\ .011 & .503 & .486 \end{pmatrix} = (.129 \quad .636 \quad .235).$$

The fixed probability vector α has the following interpretations. In the individual process it represents the long-range predictions for the occupation of an individual. Our basic theorem for regular chains tells us that these predictions are independent of the present occupational class. For the collective process, the fixed vector gives the equilibrium fractions. When these fractions are realized, the fractions in successive generations remain the same. No matter what the initial fractions are, they will, after a number of generations, be close to those given by the fixed vector.

In our example, the fixed vector is $\alpha = (.067, .624, .309)$. The actual fraction in each of the classes from the data which determined the matrix P was $\tilde{\alpha} = (.076, .634, .290)$. Thus we can see that the system may be considered to be nearly in equilibrium.

The mean first passage times have the usual interpretation for the individual process but do not seem to have a natural interpretation for the collective process. For our example, the mean first passage time matrix is

$$M = \begin{matrix} & \mathbf{U} & \mathbf{M} & \mathbf{L} \\ \mathbf{U} \\ \mathbf{M} \\ \mathbf{L} \end{matrix}\!\!\begin{pmatrix} 14.9 & 2.1 & 5.6 \\ 25.1 & 1.6 & 4.3 \\ 26.5 & 1.9 & 3.2 \end{pmatrix}.$$

The standard deviations for the first passage times are

$$\begin{matrix} & \mathbf{U} & \mathbf{M} & \mathbf{L} \\ \mathbf{U} \\ \mathbf{M} \\ \mathbf{L} \end{matrix}\!\!\begin{pmatrix} 22.5 & 1.5 & 4.1 \\ 25.0 & 1.2 & 3.9 \\ 25.1 & 1.4 & 3.5 \end{pmatrix}$$

Since the standard deviations of the first passage times are of the same order of magnitude as the means, the means are not to be taken as typical values. However, the relative size is of interest. For example, the mean time to go from lower to upper is about five times as big as the mean time to go from upper to lower.

Assume now that the individual process is in equilibrium. Then the reverse transition matrix gives the probabilities for the father's occupation when that of the son is known. If P is reversible, then, given that a man is in class t, the probability that his son will be in a given occupational class j is the same as the probability that his father was in this class j. The condition for reversibility has an interesting interpretation for the collective process. Recall that the condition for reversibility may be expressed by saying that $D^{-1}P$ should be a symmetric matrix. In other words, that $a_i p_{ij} = a_j p_{ji}$. In the collective process $a_i p_{ij}$ represents, in equilibrium, the fraction of the people in the i-th occupational class which move in one generation from the i-th to the j-th class. Also $a_j p_{ji}$ represents the fraction which move from the j-th class to the i-th class. Hence the condition for reversibility means that there should be an "equal exchange" between classes. If there is an equal exchange then clearly the total numbers in each class will remain fixed, i.e. the process will be in equilibrium. However, equal exchange is a much stronger condition. The above discussion suggests that for the collective process the matrix $D^{-1}P$ is an interesting matrix. We call this matrix the *exchange matrix*. For our basic example, the exchange matrix is

$$D^{-1}P = \begin{array}{c} \\ \mathbf{U} \\ \mathbf{M} \\ \mathbf{L} \end{array} \begin{array}{c} \begin{array}{ccc} \mathbf{U} & \mathbf{M} & \mathbf{L} \end{array} \\ \begin{pmatrix} .030 & .032 & .005 \\ .034 & .436 & .154 \\ .003 & .155 & .150 \end{pmatrix} \end{array}.$$

Note that there is approximately equal exchange between the classes.†

Finally we consider the question of lumpability for mobility processes. This is particularly important for the following reason. If we decide that the Markov assumption is reasonable for a certain method of classification, then we cannot arbitrarily treat a coarser classification as a Markov chain. This is because the coarser classification is obtained from the finer by lumping states. We know that only under very special conditions will this again result in a Markov chain. Of course the method of classification itself has a great deal to do with whether or not the Markov assumption is realistic. Hence the coarser analysis may be taken as a Markov chain even when the condition for lumpability is not satisfied. However, we must then admit that the finer analysis is not a Markov chain. We cannot have both, unless the condition for lumpability is satisfied.

We shall illustrate the above ideas in terms of some actual mobility studies. The example that we have been considering was actually obtained from a finer analysis of the data obtained by Glass and Hall for England and Wales in 1949. These authors used seven classes. They are:

1. Professional and high administrative.
2. Managerial and executive.
3. Inspectional, supervisory, and other non-manual (higher grade).

† For a more detailed discussion of the exchange properties of a system see J. Berger and J. L. Snell, "On the Concept of Equal Exchange," *Behavioral Science*, **2**, (No. 2): 111–118, 1957.

4. Same (lower grade). *
5. Skilled manual and routine grades of non-manual.
6. Semi-skilled manual.
7. Unskilled manual.

From their data we obtain the transition matrix

$$
P = \begin{array}{c c c c c c c c}
 & 1 & 2 & 3 & 4 & 5 & 6 & 7 \\
 & 0.388 & 0.147 & 0.202 & 0.062 & 0.140 & 0.047 & 0.016 \\
 & 0.107 & 0.267 & 0.227 & 0.120 & 0.207 & 0.053 & 0.020 \\
 & 0.035 & 0.101 & 0.188 & 0.191 & 0.357 & 0.067 & 0.061 \\
 & 0.021 & 0.039 & 0.112 & 0.212 & 0.431 & 0.124 & 0.062 \\
 & 0.009 & 0.024 & 0.075 & 0.123 & 0.473 & 0.171 & 0.125 \\
 & 0.000 & 0.013 & 0.041 & 0.088 & 0.391 & 0.312 & 0.155 \\
 & 0.000 & 0.008 & 0.036 & 0.083 & 0.364 & 0.235 & 0.274
\end{array}.
$$

Our previous example was obtained from this study by calling {1,2} the upper class, {3,4,5} the middle class, and {6,7} the lower class. The fixed vector is

$$\alpha = (.023 \quad .041 \quad .088 \quad .127 \quad .410 \quad .182 \quad .129).$$

The above transition matrix was estimated from a sample of 3497. The distribution of the occupations in this sample was

$$\bar{\alpha} = (.030 \quad .046 \quad .094 \quad .131 \quad .409 \quad .170 \quad .121).$$

We see that these numbers are fairly close to the equilibrium vector.

$$
M = \begin{array}{c c c c c c c c}
 & 1 & 2 & 3 & 4 & 5 & 6 & 7 \\
1 & 43.9 & 26.2 & 9.9 & 9.2 & 4.0 & 8.4 & 11.5 \\
2 & 63.1 & 24.2 & 10.1 & 8.5 & 3.5 & 8.1 & 11.1 \\
3 & 70.3 & 30.5 & 11.4 & 8.0 & 2.9 & 7.6 & 10.3 \\
4 & 72.3 & 33.0 & 12.7 & 7.9 & 2.6 & 7.0 & 10.0 \\
5 & 73.7 & 33.9 & 13.5 & 8.7 & 2.4 & 6.5 & 9.3 \\
6 & 74.9 & 34.6 & 14.1 & 9.1 & 2.6 & 5.5 & 8.8 \\
7 & 75.0 & 34.8 & 14.3 & 9.2 & 2.7 & 5.9 & 7.7
\end{array}.
$$

The diagonal entries of M are the reciprocals of the fixed vector. Since the fixed vector is close to the actual fraction of people in each class, when this fraction is small the mean time to return is correspondingly large. For a given occupational class it is interesting to compare the mean time to reach this class, starting in each of the other classes. We observe that in general the mean time to go from any state i to a given state j decreases as i gets closer to state j. Positions of these occupational classes correspond to their relative social prestige, and hence "closer" means closer in terms of prestige.

We have discussed only regular chain concepts in this section. We know that absorbing chain ideas can be fruitfully used to study regular chains. For example, we can study the behavior of the middle classes 3,4,5 by making the upper classes 1 and 2 and lower classes 6 and 7 into absorbing states. Doing this, we obtain an absorbing chain with matrices Q and R given by

$$Q = \begin{array}{c} \\ 3 \\ 4 \\ 5 \end{array} \begin{array}{c} \begin{array}{ccc} 3 & 4 & 5 \end{array} \\ \begin{pmatrix} .188 & .191 & .357 \\ .112 & .212 & .431 \\ .075 & .123 & .473 \end{pmatrix} \end{array}$$

$$R = \begin{array}{c} \\ 3 \\ 4 \\ 5 \end{array} \begin{array}{c} \begin{array}{cccc} 1 & 2 & 6 & 7 \end{array} \\ \begin{pmatrix} .035 & .101 & .067 & .061 \\ .021 & .039 & .124 & .062 \\ .009 & .024 & .171 & .125 \end{pmatrix} \end{array}.$$

The basic quantities for this chain are:

$$N = \begin{array}{c} \\ 3 \\ 4 \\ 5 \end{array} \begin{array}{c} \begin{array}{ccc} 3 & 4 & 6 \end{array} \\ \begin{pmatrix} 1.44 & .58 & 1.45 \\ .36 & 1.60 & 1.55 \\ .29 & .45 & 2.47 \end{pmatrix} \end{array} \qquad \tau = \begin{array}{c} 3 \\ 4 \\ 5 \end{array} \begin{pmatrix} 3.47 \\ 3.51 \\ 3.21 \end{pmatrix}$$

$$B = \begin{array}{c} \\ 3 \\ 4 \\ 5 \end{array} \begin{array}{c} \begin{array}{cccc} 1 & 2 & 6 & 7 \end{array} \\ \begin{pmatrix} .08 & .20 & .42 & .30 \\ .06 & .14 & .49 & .32 \\ .04 & .11 & .49 & .36 \end{pmatrix} \end{array}.$$

From τ we obtain the mean time to leave the set {3,4,5} for the first time for each starting state in the set. We see that this is between 3 and 4 for each starting state. From B we find the probabilities of leaving by moving to each of the states 1,2,6,7. Combining states

1 and 2, and 6 and 7 we can find the probability of moving out of each of the middle classes by moving to the upper or to the lower class. These probabilities are:

$$\begin{array}{c} \\ 3 \\ 4 \\ 5 \end{array}\begin{array}{c} \begin{array}{cc} \mathbf{U} & \mathbf{L} \end{array} \\ \begin{pmatrix} .28 & .72 \\ .20 & .81 \\ .15 & .85 \end{pmatrix} \end{array}$$

In each case the probability of leaving by way of the lower class is much higher than leaving by way of the upper class.

It is interesting to observe that the probability of leaving the middle class by way of the upper class decreases the lower the level of the occupational class.

As mentioned earlier, our basic example in this section was obtained by combining states in this seven-state chain. The partition used was $\mathbf{A} = (\{1,2\}, \{3,4,5\}, \{6,7\})$. The first set being the upper class, the second the middle class, and the third the lower class. It is interesting then to check the condition for lumpability with respect to this partition. To do this we must find the matrix PV (see §6.3). We obtain,

$$PV = \begin{array}{c} \\ 1 \\ 2 \\ 3 \\ 4 \\ 5 \\ 6 \\ 7 \end{array}\begin{array}{c} \begin{array}{ccc} \mathbf{A}_1 & \mathbf{A}_2 & \mathbf{A}_3 \end{array} \\ \left(\begin{array}{ccc} .534 & .404 & .062 \\ .374 & .553 & .073 \\ \hline .136 & .736 & .128 \\ .060 & .754 & .128 \\ .033 & .671 & .296 \\ \hline .013 & .520 & .467 \\ .008 & .483 & .509 \end{array}\right) \end{array}$$

To satisfy the condition for lumpability it is necessary that the components of a column of this vector be constant within the sets $\mathbf{A}_1, \mathbf{A}_2. \mathbf{A}_3$. This is certainly not the case. For example the probability of moving to $\mathbf{A}_1$ is quite different for the states of $\mathbf{A}_2$. It is .033 from state 5, .060 from state 4, and .136 from state 3. Thus we would not be justified in treating both of our processes as Markov chains.

If we choose to believe that the seven-state chain is a Markov chain, then the three-state process is not a Markov chain, but in equilibrium the matrix $\hat{P}$, the vector $\hat{a}$, and the matrix $\hat{M}$ are all well defined. If we compute these matrices by the method given in § 6.4, we obtain,

$$\hat{P} = \begin{array}{c} \\ \mathbf{U} \\ \mathbf{M} \\ \mathbf{L} \end{array} \begin{array}{c} \begin{array}{ccc} \mathbf{U} & \mathbf{M} & \mathbf{L} \end{array} \\ \begin{pmatrix} .43 & .50 & .07 \\ .05 & .70 & .25 \\ .01 & .50 & .48 \end{pmatrix} \end{array}$$

$$\hat{a} = (.06 \quad .63 \quad .31)$$

$$\hat{M} = \begin{array}{c} \\ \mathbf{U} \\ \mathbf{M} \\ \mathbf{L} \end{array} \begin{array}{c} \begin{array}{ccc} \mathbf{U} & \mathbf{M} & \mathbf{L} \end{array} \\ \begin{pmatrix} 15.7 & 2.0 & 5.9 \\ 26.3 & 1.6 & 6.3 \\ 28.0 & 2.0 & 3.0 \end{pmatrix} \end{array}$$

We see that these quantities are all quite close to those obtained by treating the three-state chain as a Markov chain.

The next example we consider is obtained from data collected by N. Rogoff† in a study made from marriage-license applications for Marion County, Indiana. The interest in this example lies in the fact that data were obtained for two different time periods, 1905 to 1912 and 1938 through the first half of 1941. Hence it is possible to compare the transition matrices for these two different time periods. Within the first sample there were 10,253 and within the second, 9,892. In the Rogoff study a very fine analysis of the occupations is made. However, for illustrative purposes, we have made a coarse analysis. This classification may be considered as non-manual, manual, and farming. We treat first the 1910 case. The transition matrix is

$$P = \begin{array}{c} \\ \text{NON-MANUAL} \\ \text{MANUAL} \\ \text{FARM} \end{array} \begin{array}{c} \begin{array}{ccc} \text{NON-MANUAL} & \text{MANUAL} & \text{FARM} \end{array} \\ \begin{pmatrix} .594 & .396 & .009 \\ .211 & .782 & .007 \\ .252 & .641 & .108 \end{pmatrix} \end{array}.$$

The fixed vector is $\alpha_{1910} = (.343 \quad .648 \quad .009)$. The actual fractions in each of the classes are given by

$$\bar{\alpha}_{1910} = (.310 \quad .658 \quad .034)$$

Note that the equilibrium vector predicts significantly fewer farmers than there actually are. This would suggest that the 1940 data should show a decrease in the fraction in farming.

† N. Rogoff, *Recent Trends in Occupational Mobility*, Glencoe, Ill., The Free Press, 1953.

For the 1940 case we find

$$P = \begin{array}{l} \text{NON-MANUAL} \\ \text{MANUAL} \\ \text{FARM} \end{array} \overset{\begin{array}{ccc} \text{NON-MANUAL} & \text{MANUAL} & \text{FARM} \end{array}}{\begin{pmatrix} .622 & .375 & .003 \\ .274 & .721 & .005 \\ .265 & .694 & .042 \end{pmatrix}}.$$

The fixed vector is $\alpha_{1940} = (.420\ .576\ .004)$. The actual fraction in in each class in 1940 is given by

$$\bar{\alpha}_{1940} = (.373 \quad .616 \quad .011).$$

As predicted, the fraction of farmers has significantly decreased. It is interesting to observe that the equilibrium vectors for 1910 and 1940 predict larger fractions in the upper class than there actually are. In the case of England the equilibrium vector predicted smaller numbers in the upper classes.

The final example that we discuss illustrates equal exchange. The data was obtained from a study made by Blumen, Kogan, and McCarthy on labor mobility.† This was a very large study based on social security records. A 1-percent sample of all workers who are or have been in covered employment since the inception of the social security system in 1937 has been kept. The study was based on a 10-percent sample from this record. It presents the following transition matrix for the group of males in the age bracket 20 to 24. We omit a discussion of the classification used in the study.

$$P = \begin{array}{c} 1 \\ 2 \\ 3 \\ 4 \\ 5 \end{array} \overset{\begin{array}{ccccc} 1 & 2 & 3 & 4 & 5 \end{array}}{\begin{pmatrix} 0.832 & 0.033 & 0.013 & 0.028 & 0.095 \\ 0.046 & 0.788 & 0.016 & 0.038 & 0.112 \\ 0.038 & 0.034 & 0.785 & 0.036 & 0.107 \\ 0.054 & 0.045 & 0.017 & 0.728 & 0.156 \\ 0.082 & 0.065 & 0.023 & 0.071 & 0.759 \end{pmatrix}}$$

The fixed vector is

$$\alpha = (.270 \quad .184 \quad .076 \quad .148 \quad .322).$$

The actual fractions in the classes considered were

$$\bar{\alpha} = (.282 \quad .170 \quad .068 \quad .137 \quad .343).$$

† I. Blumen, M. Kogan, P. J. McCarthy, *The Industrial Mobility of Labor as a Probability Process*, Cornell Studies in Industrial and Labor Relations, Vol. VII, 1955.

The most interesting feature of this example is the exchange matrix. This is

$$D^{-1}P = \begin{matrix} 1 \\ 2 \\ 3 \\ 4 \\ 5 \end{matrix} \begin{pmatrix} .225 & .009 & .004 & .008 & .026 \\ .008 & .145 & .003 & .007 & .021 \\ .003 & .002 & .060 & .003 & .008 \\ .008 & .007 & .002 & .108 & .023 \\ .026 & .021 & .007 & .023 & .244 \end{pmatrix}.$$

The almost perfect symmetry of this matrix indicates that this system may be considered to be in equal exchange in equilibrium, or that the process is reversible.

§ 7.7 **The open Leontief model.** In the Leontief input-output model, we consider an economy in which there are r industries and we make the simplifying assumption that each industry produces exactly one kind of goods. We regard the natural factors of production such as land, timber, minerals, etc. as free, and do not consider them as entering into the cost of finished goods. In general, the industries are interconnected in the sense that each must buy a certain amount (positive or zero) of the other's products in order to run its industry. We shall define *technological coefficients* as follows: q_{ij} is the amount of the output of industry j that must be purchased by industry i in order that industry i may produce \$1 worth of its own goods. Let Q be the $r \times r$ matrix with entries q_{ij}. By their definition, the technological coefficients are non-negative.

It is easy to see that the sum of the q_{ij}, for i fixed, gives the total value of the inputs needed by the i-th industry in order to produce \$1 worth of its goods. If the i-th industry is to be profitable, or at least to break even, this sum must be less than or equal to the value of its output, i.e. $q_{i1}+q_{i2}+ \cdots +q_{ir} \leqslant 1$. For obvious reasons we shall call the i-th industry *profitable* if the strict inequality holds and *profitless* if the equality holds. We make the assumption that every industry is either profitable or profitless and thus rule out the possibility of unprofitable industries.

We can restate the above conditions as

$$Q \geqslant 0 \tag{1}$$

$$Q\xi \leqslant \xi. \tag{2}$$

Having discussed the inputs of the industries we next discuss their outputs. Let x_i denote the monetary value of the output of the i-th industry and let $\pi = (x_1, x_2, \ldots, x_r)$ be the row vector of outputs. Since the i-th industry needs an amount $x_i q_{ij}$ of the output of the j-th industry, the vector of inputs needed by the industries is simply πQ. Then the j-th component of πQ gives the total value of the output that must be produced by the j-th industry in order to meet the inter-industry demand for its product.

Let us assume that the economy supplies for consumption an amount c_i of the output of the i-th industry. Let $\gamma = (c_1, c_2, \ldots, c_r)$ be the consumption vector; we shall require that

$$\gamma \geqslant 0. \tag{3}$$

The requirement that the production vector of the economy be adjusted so that the inter-industry needs as well as the consumption needs may be fulfilled is now easy to write in vector form; it is

$$\pi = \pi Q + \gamma. \tag{4}$$

Rewriting (4) as

$$\pi(I - Q) = \gamma, \tag{5}$$

we see that it is a set of r simultaneous equations in r unknowns.

To be economically meaningful, we must find non-negative solutions to (5). Since the demand vector γ may be arbitrary, we see that equations (5) are in general non-homogeneous and will have a solution if and only if the matrix $I - Q$ has an inverse. Moreover, the solutions to (5) will be non-negative for *every* γ if and only if $(I - Q)^{-1}$ has all non-negative components. We must therefore search for necessary and sufficient conditions that the inverse of $I - Q$ be non-negative.

We will solve this problem by imbedding our model in a Markov chain. (This solution was worked out by the authors jointly with G. L. Thompson.)

By the *Markov chain associated with an input-output model* we shall mean a Markov chain with the following properties:

(i) The states are the r processes of the model plus one additional absorbing state s_0, called the *banking state*.
(ii) The transition matrix P is defined as follows:

$$\begin{aligned} p_{00} &= 1 & & \\ p_{0j} &= 0 & j &> 0 \\ p_{ij} &= q_{ij} & i,j &> 0 \\ p_{i0} &= 1 - \sum_{j=1}^{r} q_{ij} & i &> 0. \end{aligned}$$

The intuitive interpretation of this is the following: If industry i receives a dollar for its use, then it spends it by buying p_{ij} from industry j. The remainder of the dollar, if any—that is, the amount p_{i0}—is the profit, and we may think of it as being deposited in a bank. The fact that the banking state is an absorbing state, means that the bank gets money but does not spend it.

We immediately see that if Q is a matrix satisfying (1) and (2), then a non-negative solution to equations (5) exists for every $r \geqslant 0$, if and only if the associated Markov chain is absorbing, with the banking state s_0 as its only absorbing state. If s_0 is the only absorbing state for an absorbing chain, then $(I-Q)^{-1}=N$ exists and is non-negative; hence $\pi=\gamma N$ is the desired solution. Otherwise $(I-Q)^{-1}$, which gives the mean number of times in various states before reaching s_0, would have to have infinite entries, i.e. cannot exist.

There is a simple economic interpretation of this result. We see that from every state it must be possible to "reach" the banking state. Only a profitable industry "reaches" the bank directly. A profitless industry must reach the bank through a profitable one. Hence our condition states that *every industry must be either profitable or must depend on a profitable industry.* For example, if we assume that every industry depends on labor, and that labor is a profitable industry (which presumably means that labor is paid more than subsistence wages), then our condition will be met, and all demands can be fulfilled.

If the above condition is violated, then the economy cannot fulfill all possible demands. Let us ask what kinds of demands it can fulfill. First of all we consider the case that there is no profitable industry. This means that each industry needs all it produces to pay for raw materials, and it would seem that it could meet no outside demand. That this is indeed the case is easily proved.

If there is no profitable industry, then each row of Q has row sum 1, hence $Q\xi=\xi$. If we multiply equation (4) by ξ on the right, we find that

$$\pi\xi = \pi Q\xi+\gamma\xi = \pi\xi+\gamma\xi,$$

hence $\gamma\xi=0$. This says that the sum of all the demands is 0. Hence no (positive) demand can be fulfilled.

Let us now consider the general case where our condition is violated. The associated Markov chain is not an absorbing chain with the single absorbing state. Then there must be an ergodic set other than $\{s_0\}$, i.e. a closed group of industries none of which is profitable, and which depend on no industry outside the group. Let us take the set of all such industries, that is the union of all ergodic sets other than $\{s_0\}$. The submatrix $\bar{Q}$ of these industries has the property $\bar{Q}\xi=\xi$, as above, and hence can fulfill no outside demand. Thus the entire economy can fulfill no demands of goods produced by these industries. And any goods whose production required raw materials from these industries also cannot be supplied, since these would act as outside demands on the closed group of industries.

However, if we remove this closed group of profitless industries and all industries depending on them, the remaining industries (if any) will fulfill our requirement, and hence can satisfy arbitrary demands. These results can be summarized: If there are industries which depend on no profitable industry, then these cannot fulfill an outside demand, and neither can any industry depending on them. The remaining industries can fulfill *any* outside demand. In terms of states this means that any ergodic state (other than s_0), or any transient state from which such a state is reachable, can fulfill no demand.

To find what industries can support a demand, we use the following simple algorithm for the classification of states:

(a) Make a check opposite each row of Q whose row-sum is less than 1; that is, check each row corresponding to a profitable industry.
(b) Check the columns having the same indices as the rows already marked and then check, in these columns, rows which have positive entries.
(c) Iterate (b) until it produces no new rows. Then one of two possibilities may occur:

   (1) All rows are checked.
   (2) Not all rows are checked.

If case (c1) occurs then the associated Markov chain is absorbing with the single absorbing state s_0. Hence any non-negative demand can be met. If case (c2) occurs, then the rows which are not checked correspond to the maximal profitless closed group. We can find all states depending on these by marking these rows (removing previous check marks) and applying (b) repeatedly. Any state so marked will not be able to fulfill an outside demand.

We thus see that the entire question of what outside demands can be met by the economy is settled by a very simple algorithm. The "computation" requires finding the row-sums, and then simple iterations in which only the positivity of components is checked. This algorithm is practical even for very large matrices.

The industries which can meet no outside demand form a totally useless segment of the economy. From here on we will assume that they have been deleted; then $(I-Q)^{-1}$ exists.

Next we want to raise the question: If an order for one dollar's worth of good is given by a customer to industry i, how much of it ends up in the hands of the various industries?

First of all we must ask what the total demand is on the various industries. We have a γ vector whose i-th component is 1, and which has 0's as other components. Hence $\pi = \gamma N$ is simply the i-th row of N. This gives a direct interpretation to the entries of N; n_{ij} is the amount industry j must produce to fill a dollar order for industry i. Since industry j makes p_{j0} profit on a unit production, the answer to our question is that if industry i is given a dollar, the profit of industry j will be $n_{ij}p_{j0}$.

The sum of all the profits is $\sum_j n_{ij}p_{j0} = b_{i0} = 1$ (since s_0 is the only absorbing state). This shows that a dollar spent by the consumer ends up as profit in the hands of the profit-making industries.

A related question is the following: If a dollar order is given to industry i, how much activity does this result in? It will result in n_{ij} units of production in industry j. The sum of these is t_i, the i-th component of τ. This is normally much greater than 1. For an order γ, the total production is $\gamma N \xi = \gamma \tau$.

Let us consider an example. Suppose that the technological coefficients for six industries are given by

$$Q = \begin{pmatrix} 1/2 & 0 & 1/4 & 0 & 0 & 0 \\ 1/4 & 1/4 & 1/4 & 0 & 0 & 0 \\ 1/2 & 0 & 1/2 & 0 & 0 & 0 \\ 0 & 0 & 0 & 1/4 & 3/4 & 0 \\ 0 & 0 & 0 & 1 & 0 & 0 \\ 0 & 1/4 & 0 & 1/4 & 0 & 1/4 \end{pmatrix},$$

then

$$P = \begin{pmatrix} 1 & 0 & 0 & 0 & 0 & 0 & 0 \\ 1/4 & 1/2 & 0 & 1/4 & 0 & 0 & 0 \\ 1/4 & 1/4 & 1/4 & 1/4 & 0 & 0 & 0 \\ 0 & 1/2 & 0 & 1/2 & 0 & 0 & 0 \\ 0 & 0 & 0 & 0 & 1/4 & 3/4 & 0 \\ 0 & 0 & 0 & 0 & 1 & 0 & 0 \\ 1/4 & 0 & 1/4 & 0 & 1/4 & 0 & 1/4 \end{pmatrix} \begin{matrix} s_0 \\ s_1 \\ s_2 \\ s_3. \\ s_4 \\ s_5 \\ s_6 \end{matrix}$$

From the first column we see that s_1, s_2, and s_6 are the profitable industries. The classification of states is given by Figure 7-9.

Here $\{s_4, s_5\}$ is the ergodic set of industries which are profitless and do not depend on profitable industries. Industry s_6 is profitable, but it depends on the former. Hence s_4, s_5, s_6 are useless, and may be

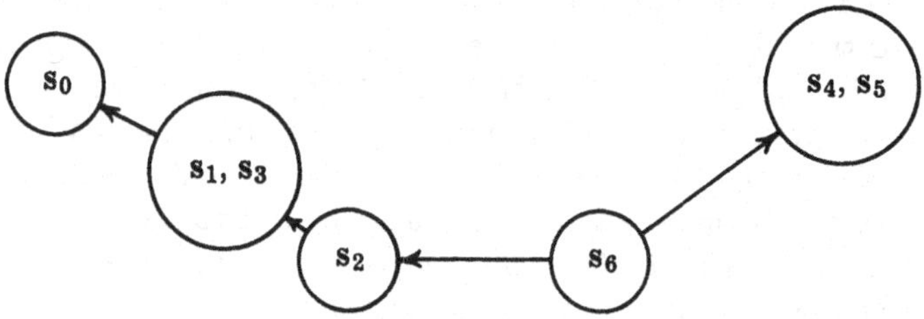

FIGURE 7-9

deleted. Industry s_3 is profitless but not useless. The deleted transition matrix is

$$P = \begin{pmatrix} 1 & 0 & 0 & 0 \\ 1/4 & 1/2 & 0 & 1/4 \\ 1/4 & 1/4 & 1/4 & 1/4 \\ 0 & 1/2 & 0 & 1/2 \end{pmatrix} \begin{matrix} s_0 \\ s_1 \\ s_2 \\ s_3 \end{matrix}.$$

Hence

$$N = \begin{pmatrix} 4 & 0 & 2 \\ 8/3 & 4/3 & 2 \\ 4 & 0 & 4 \end{pmatrix} \qquad \tau = \begin{pmatrix} 6 \\ 6 \\ 8 \end{pmatrix}.$$

Thus, for example, a unit order to s_2 will stimulate a total of 6 dollars' production, $8/3$ from s_1, $4/3$ from s_2, and 2 from s_3. On this dollar order s_1 makes a profit of $8/3 \cdot 1/4 = 2/3$, s_2 makes $4/3 \cdot 1/4 = 1/3$, and s_3 makes $2.0 = 0$ (s_3 is profitless).

If we place an outside demand of $\gamma = (1,3,2)$ on the economy, then $\gamma N = (20,4,16)$ units will have to be produced by the various industries. The total production is worth $\gamma\tau = 40$ dollars.

We note that P is lumpable into the partition $[\{s_0\}, \{s_1, s_2\}, \{s_3\}]$. The lumped process yields

$$\hat{P} = \begin{pmatrix} 1 & 0 & 0 \\ 1/4 & 1/2 & 1/4 \\ 0 & 1/2 & 1/2 \end{pmatrix} \qquad \hat{N} = \begin{pmatrix} 4 & 2 \\ 4 & 4 \end{pmatrix} \qquad \hat{\tau} = \begin{pmatrix} 6 \\ 8 \end{pmatrix}.$$

For

$$\hat{\gamma} = (4,2) \qquad \hat{\gamma}\hat{N} = (24,16) \qquad \hat{\gamma}\hat{\tau} = 40.$$

The "lumped" industries s_1 and s_2 are here considered as acting as one "industrial group." This process will yield the total demands on the industrial group, but not the breakdown into industrial demands.

For practical problems the computations may be prohibitive. Hence we are often happy to solve a lumped version of the economy. The condition for lumpability becomes the following: Any industry in one industrial group makes the same total per unit demands on the members of (its own or) another industrial group. (Then all industries in the group make the same per unit profit.) While these conditions are unlikely to be met exactly, they may be put to a good approximation. This allows us to take industrial groups as basic entities, and yields a smaller and more manageable model.

We thus see that even in a non-probabilistic model a great deal of information can be obtained from Markov chain theory.

APPENDICES

I—Summary of Basic Notation

Page

25 $\mathbf{M}_i[\mathbf{f}]$, $\mathbf{Var}_i[\mathbf{f}]$, $\mathbf{Pr}_i[\mathbf{p}]$ denote the mean value of the function **f**, variance of **f**, and probability of the statement **p** when the chain is started in state $\mathbf{s}_i$.

19 $R = \{r_{ij}\}$ matrix with entries r_{ij}

19 $\rho = \{r_j\}$ row vector with component r_j

19 $\gamma = \{c_i\}$ column vector with component c_i

20 ξ column vector with all entries 1

20 η row vector with all entries 1

20 E matrix with all entries 1

20 I identity matrix

20 O matrix with all entries 0

46 $\mathbf{d}_{ij} = \begin{cases} 1 \text{ if } i = j \\ 0 \text{ if } i \neq j \end{cases}$

21 A^T is the transpose of A

21 A_{sq} results from A by squaring each entry

21 A_{dg} results from A by setting off-diagonal entries equal to 0

II—Basic Definitions

25 A *finite Markov chain* is a stochastic process which moves through a finite number of states, and for which the probability of entering a certain state depends only on the last state occupied.

35 An *ergodic set* of states is a set in which every state can be reached from every other state, and which cannot be left once it is entered.

35 A *transient set* of states is a set in which every state can be reached from every other state, and which can be left.

35 An *ergodic state* is an element of an ergodic set.

Page

35 A *transient state* is an element of a transient set.

35 An *absorbing state* is a state which once entered is never left.

37 An *absorbing chain* is one all of whose ergodic states are absorbing; or—equivalently—which has at least one absorbing state, and such that an absorbing state can be reached from every state.

37 An *ergodic chain* is one whose states form a single ergodic set; or—equivalently—a chain in which it is possible to go from every state to every other state.

37 A *cyclic chain* is an ergodic chain in which each state can only be entered at certain periodic intervals.

37 A *regular chain* is an ergodic chain that is not cyclic.

III—Basic Quantities for Absorbing Chains

STANDARD FORM FOR TRANSITION MATRIX

44
$$P = \left(\begin{array}{c|c} I & O \\ \hline R & Q \end{array}\right)$$

46 $\mathbf{n}_j$ number of times in state $\mathbf{s}_j$ before absorption

46 $\mathbf{u}^k{}_j$ function that is 1 if process is in $\mathbf{s}_j$ after k steps, 0 otherwise

50 $\mathbf{t}$ number of steps taken before absorption

52 b_{ij} probability starting in state $\mathbf{s}_i$ that the process is absorbed in state $\mathbf{s}_j$

61 $\mathbf{r}_i$ number of times in transient state $\mathbf{s}_i$ before leaving the state

61 h_{ij} probability starting in $\mathbf{s}_i$ that the process is ever in $\mathbf{s}_j$

61 $\mathbf{m}$ total number of transient states entered before absorption

46 $N = \{\mathbf{M}_i[\mathbf{n}_j]\}$

49 $N_2 = \{\mathbf{Var}_i[\mathbf{n}_j]\}$

49 $\tau = \{\mathbf{M}_i[\mathbf{t}]\}$

49 $\tau_2 = \{\mathbf{Var}_i[\mathbf{t}]\}$

49 $B = \{\mathbf{b}_{ij}\}$

62 $\mu = \{\mathbf{M}_i[\mathbf{m}]\}$

63 $\hat{P}$ transition matrix for process of changes of state

IV—Basic Formulas for Absorbing Chains

Page	
46	$N = (I - Q)^{-1}$ (Fundamental matrix)
49	$N_2 = N(2N_{dg} - I) - N_{sq}$
49	$B = NR$
61	$H = [N - I]N_{dg}^{-1}$
49	$\tau = N\xi$
49	$\tau_2 = (2N - I)\tau - \tau_{sq}$
61	$\mathbf{M}_i[\mathbf{r}_i] = \dfrac{1}{1 - p_{ii}}$
61	$\mathbf{Var}_i[\mathbf{r}_i] = \dfrac{p_{ii}}{(1 - p_{ii})^2}$
62	$\mu = (NN_{dg}^{-1})\xi$

V—Basic Quantities for Ergodic Chains

70	$\alpha = \{a_i\}$ fixed probability vector for P
88	$c_{ij} = \lim_{n \to \infty} \frac{1}{n} \mathbf{Cov}_k[\mathbf{y}_i^{(n)}, \mathbf{y}_j^{(n)}]$
88	$\beta = \{b_j\}$ vector of limiting variances for the number of times in each of the states $a_j(2z_{jj} - a_j)$ [see p. 91]
70	A matrix with each row α
79	D diagonal matrix with j-th entry $1/a_j$
105	$\hat{P}$ transition matrix for the reverse process
75	$Z = (I - P + A)^{-1}$ (Fundamental matrix)
81	$c = \sum_i z_{ii}$
78	$M = \{m_{ij}\}$ matrix of mean number of steps required to reach $\mathbf{s}_j$ for the first time, starting in $\mathbf{s}_i$
82	$W = \{w_{ij}\}$ matrix of variances for the number of steps required to reach $\mathbf{s}_j$, starting in state $\mathbf{s}_i$
76	$\mathbf{y}_j^{(n)}$ number of times in state $\mathbf{s}_j$ in the first n steps

VI—Basic Formulas for Ergodic Chains

Page	
105	$\hat{P} = DP^T D^{-1}$
78	$M = \{m_{ij}\} = (I - Z + EZ_{dg})D$
81	$\bar{M} = M - M_{dg}$
83	$W = \{w_{ij}\} = M(2Z_{dg}D - I) + 2(ZM - E(ZM)_{dg})$
85	$C = \{c_{ij}\} = \{a_i z_{ij} + a_j z_{ji} - a_i d_{ij} - a_i a_j\}$
79	$m_{ii} = \dfrac{1}{a_i}$
83	$w_{ii} = \dfrac{2z_{ii}}{a_i^2} - \dfrac{1}{a_i}$
81	$\alpha M = \eta Z_{dg} D$
81	$M\alpha^T = c\xi$
81	$\alpha = (c-1)[(\bar{M})^{-1}\xi]^T$
81	$P = I + (D - E)(\bar{M})^{-1}$

VII—Some Basic Examples

Page	Example
47	*1a*
103	*2a*
90	*3a*
108	*3a*
48	*10a*
102	*11a*
42	*13*
42	*14*

VIII—Generalization of a Fundamental Matrix

John G. Kemeny

ABSTRACT

It is shown that, for a finite ergodic Markov chain, basic descriptive quantities, such as the stationary vector and mean first-passage matrix, may be calculated using any one of a class of fundamental matrices. New applications of the use of these operators are discussed.

INTRODUCTION

The motivation for this paper is to generalize the concept of the fundamental matrix of a finite ergodic Markov chain (see [4]). The approach will be to consider a class of linear equations, of which the Markov chain case is a subclass. It will be shown that one gains new insight into previous work on ergodic chains, and that the generalized operator has interesting new applications.

We assume that we are dealing with an n-dimensional vector space, for some fixed $n>1$. Unless otherwise indicated, in our formulas capital letters will denote n-by-n matrices, lowercase letters denote n-component column vectors, and Greek letters denote n-component row vectors.

A CLASS OF LINEAR EQUATIONS

A standard approach to the computation of key quantities for finite Markov chains is to compare the value of an unknown quantity x with its expected value after one step. This leads to an equation of the form

$$x=Px+f, \tag{1}$$

where P is the transition matrix and the components of f are known quantities. Let us write this class of equations in a standard form:

$$(I-P)x=f. \tag{2}$$

We wish to consider equations of this form in general, using the special case of Markov chains only as motivation. It is well known that the nature of the solutions of (2) is determined by studying the homogeneous equation

$$(I-P)x=0. \tag{3}$$

If this equation has only the solution $x=0$, then (2) has a unique solution: if there is a nonzero solution of (3), then (2) has either no solution or infinitely many solutions, depending on the nature of f.

The case where (3) has only the trivial solution is the easy case. Then $I-P$ is nonsingular, and the solution of (2) is $x=(I-P)^{-1}f$. This is the case for a finite transient (absorbing) Markov chain, where P is the transition matrix restricted to transient states. Therefore, the computation of basic quantities for such chains is simple (see [4], Chapter 3).

The case we wish to consider is one in which $I-P$ is singular. Specifically, we shall assume that it has a one-dimensional kernel. (This is always the case for finite ergodic chains.)

ASSUMPTION. The homogeneous equation (3) has a nonzero solution $x=h$, and every solution of (3) is a multiple of h.

It should be noted that h is a fixed point of the transformation P, i.e., $Ph=h$. From linear algebra we know that there is also a fixed point of P acting on row vectors, $\alpha P=\alpha$. From now on we shall use h and α for these fixed points. It should be remembered that they are determined only up to a constant multiple. In the case of an ergodic chain, h may be chosen as the vector all of whose components are 1, and α as the all-positive probability vector of limiting probabilities.

We shall next demonstrate that there is a simple modification of the matrix $I-P$ which is nonsingular, and which may be used to solve (2). The modification allows us to choose row and column vectors β and g almost arbitrarily, giving a great deal of flexibility to the generalization. The reader should keep in mind that while the product βg is a number, the product $g\beta$ is an n-by-n matrix.

THEOREM 1. *Let β and g be any two vectors such that βh and αg are nonzero. Then the inverse*

$$Z=(I-P+g\beta)^{-1} \tag{4}$$

exists.

Proof. Suppose that

$$(I-P+g\beta)x=0. \tag{5}$$

Then

$$x=Px-g(\beta x). \tag{6}$$

We multiply the equation by α, and use the fact that $\alpha P=\alpha$. This yields $(\alpha g)(\beta x)=0$. But $\alpha g\neq 0$; hence $\beta x=0$. Thus (6) reduces to $x=Px$. By our Assumption, $x=ch$, where c is a constant. Then $0=\beta x=c(\beta h)$, and $\beta h\neq 0$; hence $c=0$. Thus (5) has only the solution $x=0$, and hence the matrix is nonsingular. ■

Let us next derive some properties of Z. From (4),

$$Z(I-P+g\beta)=I. \tag{7}$$

if we multiply this equation on the right by h, and use $Ph=h$, we obtain $(Zg)(\beta h)=h$, or

$$Zg=\frac{1}{\beta h}h. \tag{8}$$

In exactly the same manner, using that Z is a right inverse, we obtain

$$\beta Z=\frac{1}{\alpha g}\alpha. \tag{9}$$

Thus from Z we can obtain the fixed points h and α. Substituting (8) into (7), we find that

$$Z(I-P)=I-\frac{1}{\beta h}h\beta, \tag{10}$$

and dually,

$$(I-P)Z=I-\frac{1}{\alpha g}g\alpha. \tag{11}$$

We are now ready to demonstrate the usefulness of Z in solving the equation (2).

THEOREM 2. *The equation* (2) *has a solution if and only if* $\alpha f=0$. *If a solution exists, one may specify the value of* βx *arbitrarily, say* $\beta x=c$ (c *a constant*), *and one obtains the unique solution*

$$x=Zf+\frac{c}{\beta h}h. \tag{12}$$

Proof. Multiplying (2) by α shows that $\alpha f=0$ is a consequence; hence this is a necessary condition for the existence of a solution. Let us next multiply (2) by Z and make use of (10). We obtain

$$x-\frac{1}{\beta h}h(\beta x)=Zf.$$

And if we impose the condition $\beta x=c$, then we see that (12) is the only possible solution. Conversely, if we substitute (12) into (2) and use (11), we find that x satisfies the equation. (Recall that $\alpha f=0$ and $Ph=h$.) And if we multiply (12) by β and use (9), we verify that the solution also satisfies the condition $\beta x=c$. ∎

We have thus shown both the necessary and sufficient condition for the existence of a solution and precisely how much additional may be required of a solution. Since β may be any row vector such that $\beta h\neq 0$, we have a very flexible tool. And the choice of β in Z is naturally determined by the nature of the side condition $\beta x=c$.

There is a dual result which shows the role of g. Its proof exactly parallels the proof of Theorem 2:

THEOREM 3. *The equation*

$$\xi(I-P)=\phi \tag{13}$$

has a solution if and only if $\phi h=0$. If a solution exists, one may specify the value ξg arbitrarily, say $\xi g=c$, and one obtains the unique solution

$$\xi=\phi Z+\frac{c}{\alpha g}\alpha. \tag{14}$$

A SPECIAL CASE

Assume that $\alpha h\neq 0$. Then we may choose them so that $\alpha h=1$. Let $\beta=\alpha$ and $g=h$. All our conditions are met, and $\beta h=\alpha g=1$. For an ergodic Markov chain the resulting operator

$$Z^*=(I-P+h\alpha)^{-1}, \qquad \alpha h=1, \tag{15}$$

is called the "fundamental matrix." For it (8) and (9) take the simple form

$$Z^*h=h \quad \text{and} \quad \alpha Z^*=\alpha, \tag{16}$$

and (12) and (14) take the forms

$$x=Z^*f+ch \qquad \text{if} \quad \alpha f=0 \text{ and } \alpha x=c. \tag{17}$$

$$\xi=\phi Z^*+c\alpha \qquad \text{if} \quad \phi h=0 \text{ and } \xi h=c. \tag{18}$$

We shall next show that any Z may be expressed in terms of Z^*, and conversely. It will be convenient to introduce the constants

$$c_1=\frac{1}{\alpha g}, \quad c_2=\frac{1}{\beta h}, \quad c_3=\beta Z^*g+1, \quad c_4=\alpha Zh+1.$$

From (10),

$$Z(I-P+h\alpha)=I-c_2 h\beta+(Zh)\alpha;$$

hence, multiplying by Z*,

$$Z = Z^* - c_2 h(\beta Z^*) + (Zh)\alpha. \tag{19}$$

Similarly,

$$Z^*(I - P + g\beta) = I - h\alpha + (Z^*g)\beta,$$

$$Z^* = Z - h(\alpha Z) + c_1(Z^*g)\alpha. \tag{20}$$

Substituting (20) into (19),

$$h[\alpha Z + c_2 \beta Z^*] = [Zh + c_1 Z^* g]\,\alpha. \tag{21}$$

Multiplying on the right by g,

$$c_2 c_3 h = \frac{1}{c_1} Zh + Z^* g.$$

We solve this for Zh and substitute in (19):

$$Z = Z^* - c_2 h(\beta Z^*) - c_1(Z^*g)\alpha + c_1 c_2 c_3 h\alpha. \tag{22}$$

This expresses Z in terms of Z*. If we multiply (21) by h and solve for Z^*g, we have from (20)

$$Z^* = Z - h(\alpha Z) - (Zh)\alpha + c_4 h\alpha, \tag{23}$$

which expresses Z* in terms of Z. Computing αZh from (22), we obtain a useful identity:

$$c_4 = c_1 c_2 c_3. \tag{24}$$

A numerical example may be helpful at this stage:

$$P = \begin{pmatrix} 2 & 1 \\ -2 & -1 \end{pmatrix}, \qquad h = \begin{pmatrix} 1 \\ -1 \end{pmatrix}, \qquad \alpha = (2, 1).$$

This satisfies all our conditions, including $\alpha h = 1$. We find that Z* is the identity matrix. (This will always be the case when $P=h\alpha$.) Suppose that we choose

$$\beta=(1,0), \qquad g=\begin{pmatrix}1\\1\end{pmatrix}.$$

Then

$$Z=\begin{pmatrix}\frac{2}{3} & \frac{1}{3}\\ -1 & 0\end{pmatrix},$$

$$c_1=\tfrac{1}{3}, \quad c_2=1, \quad c_3=2, \quad c_4=\tfrac{2}{3},$$

and the previous results are easily verified. We wish to solve the equation

$$(I-P)x=\begin{pmatrix}2\\-4\end{pmatrix}.$$

Since $\alpha f=0$, it does have solutions. We may specify βx, i.e., the first component of x. If we require $\beta x=3$,

$$x=Z\begin{pmatrix}2\\-4\end{pmatrix}+3h=\begin{pmatrix}3\\-5\end{pmatrix},$$

which satisfies all our requirements.

It should be pointed out that, while Z^* always exists for an ergodic chain, there are cases where Theorem 1 applies but Z^* does not exist. A simple example will illustrate this: let

$$P=\begin{pmatrix}1 & 0\\ 1 & 1\end{pmatrix}, \qquad \alpha=(1,0), \qquad h=\begin{pmatrix}0\\1\end{pmatrix}.$$

All the assumptions of the first section are met, and hence the matrices (4) exist and have the stated properties. But, since $\alpha h=0$, Z^* is not one of them. Indeed, $I-P+h\alpha=0$, and certainly does not have an inverse. This shows that the method of this paper provides not only added flexibility but also wider applicability than Z^*.

ERGODIC CHAINS

A finite Markov chain is ergodic if from any state it is possible to reach every other state. If P is the transition matrix of such a chain, and h is the constant vector (all components equal to 1), then the Assumption of this paper is always satisfied. Such a Markov chain has an equilibrium, i.e., a probability vector α such that $\alpha P=\alpha$. And α is strictly positive. Thus there are natural fixed points h and α, and $\alpha h=1$.

The matrix Z^* of the previous section is called the *fundamental matrix* of the ergodic chain, and it can be shown that the various interesting probabilistic quantities can be computed in terms of α and Z^*. In particular, Eqs. (16), (17), and (18) are well known results about finite ergodic chains (See [4].) Two other important results are that the mean time to return to a state j is

$$m_j=1/\alpha_j \tag{25}$$

and the mean time to go from i to j (mean first-passage time) is

$$M_{ij}=\frac{Z^*_{jj}-Z^*_{ij}}{\alpha_j}. \tag{26}$$

To have a numerical illustration available, we introduce the weather in the Land of Oz (see [3]):

$$P=\begin{pmatrix} \frac{1}{2} & \frac{1}{4} & \frac{1}{4} \\ \frac{1}{2} & 0 & \frac{1}{2} \\ \frac{1}{4} & \frac{1}{4} & \frac{1}{2} \end{pmatrix}, \qquad h=\begin{pmatrix} 1 \\ 1 \\ 1 \end{pmatrix}, \qquad \alpha=\left(\tfrac{2}{5},\tfrac{1}{5},\tfrac{2}{5}\right),$$

$$Z^*=\frac{1}{75}\begin{pmatrix} 86 & 3 & -14 \\ 6 & 63 & 6 \\ -14 & 3 & 86 \end{pmatrix}, \qquad M=\begin{pmatrix} 0 & 4 & \frac{10}{3} \\ \frac{8}{3} & 0 & \frac{8}{3} \\ \frac{10}{3} & 4 & 0 \end{pmatrix},$$

$$m=\left(\tfrac{5}{2},5,\tfrac{5}{2}\right).$$

This treatment of finite ergodic chains has never seemed as satisfactory as the treatment of finite transient chains. For the latter $(I-Q)^{-1}$ is the natural fundamental matrix, where Q represents transitions from transient state to transient state, and all quantities can be expressed in terms of it. For ergodic chains Z^* appears somewhat arbitrary. It also suffers from the difficulty that one must compute α (solving n equations) before one can compute Z^*. Various alternatives to this matrix have since appeared in the literature, (see Meyer [7, 8]).

We now know that Z^* is only one of an infinite number of possible choices for the fundamental matrix. One may choose any Z, the only restrictions are that αg and βh should not be 0. Thus β may be any vector such that the sum of the components is not zero, and if g is chosen as a nonnegative and nonzero vector, then $\alpha g \neq 0$—irrespective of what α may be.

Let us propose a new approach to the treatment of finite ergodic chains. Let

$$Z_\beta = (I - P + h\beta)^{-1}, \qquad \beta h = 1. \tag{27}$$

That is, we let $g = h$, and β be any vector with row sum 1. Theorem 1 guarantees its existence. Then, from (8) and (9),

$$Z_\beta h = h \quad \text{and} \quad \beta Z_\beta = \alpha. \tag{28}$$

Thus we may find α from the fundamental matrix rather than having to find α first. Other quantities are determined from an equation of the form (1), with $\alpha f = 0$; we know from Theorem 2 that we may impose the additional condition $\beta x = c$ and obtain the unique solution

$$x = Z_\beta f + ch. \tag{29}$$

How are the mean first-passage times expressed in terms of the generalized fundamental matrix? Instead of retracing the derivation of the matrix M, we use (23) to translate the formula (26):

$$Z^*_{jj} - Z^*_{ij} = (Z_{jj} - Z_{ij}) - (h_j - h_i)(\alpha Z)_j - (Zh)_j \alpha_j + (Zh)_i \alpha_j + c_4(h_j - h_i)\alpha_j.$$

And since h is a constant vector, $h_j - h_i = 0$:

$$M_{ij} = \frac{Z_{jj} - Z_{ij}}{\alpha_j} - (Zh)_j + (Zh)_i. \tag{30}$$

If we use as Z a Z_β, then (28) shows that the last two terms cancel. Thus the simple formula (26) holds for any Z_β in place of Z^*.

This can be seen more simply if we express Z_β in terms of Z^*. From (22), using $g = h$, $\beta h = \alpha h = 1$, and (16),

$$Z_\beta = Z^* - h(\beta Z^* - \alpha). \tag{31}$$

For the Oz example, if we select $\beta=(\frac{1}{2},0,\frac{1}{2})$, then

$$Z_\beta=\begin{pmatrix} \frac{16}{15} & \frac{1}{5} & -\frac{4}{15} \\ 0 & 1 & 0 \\ -\frac{4}{15} & \frac{1}{5} & \frac{16}{15} \end{pmatrix}.$$

It is easy to verify that $Z_\beta h=h$ and $\beta Z_\beta=\alpha$, and that M is given by (26) using Z_β in place of Z^*. We can also verify (31) by direct computation.

Thus any Z_β is a suitable fundamental matrix, and it can be computed without knowing α.

APPLICATIONS

Consider a Markov chain that has the property that when it moves away from a state i, it moves to any other state with the same probability p_i. For example, Oz has this property with $p_1=p_3=\frac{1}{4}$ and $p_2=\frac{1}{2}$. If we choose $g_i=p_i$ and $\beta_i=1$, then $I-P+g\beta$ is the diagonal matrix with entries np_i. Thus Z is diagonal matrix with $Z_{ii}=1/(np_i)$. From (9) we know that α is proportional to βZ. Therefore,

$$\alpha_i=\frac{1}{p_i}\div\sum_k\frac{1}{p_k}, \tag{32}$$

and from (30),

$$M_{ij}=\frac{1}{n}\left(\sum_k\frac{1}{p_k}-\frac{1}{p_j}+\frac{1}{p_i}\right). \tag{33}$$

For Oz, $\Sigma_k 1/p_k=10$. Thus, for example $\alpha_1=\frac{4}{10}=\frac{2}{5}$ and $M_{12}=\frac{1}{3}(10-2+4)$ $=4$.

Next we consider the method used in [6] to compute α. The "recipe" is to replace the last column of $I-P$ by ones and invert; then α is the last row of the inverse. This corresponds to choosing $g_i=1-(I-P)_{in}$ and $\beta=(0,\ldots,0,1)$. The inverse in question is the Z corresponding to this choice of g and β, and $\alpha g=1$. Hence from (9), $\alpha=\beta Z$, which is the last row of Z. Meyer [7] showed that this matrix could be used as a fundamental matrix in place of Z^*.

A third interesting choice for g and β is the following. We choose as β the first row of P, and $g_1=1$ while the other components of g are 0. Then Z has the form

$$Z=\left(\begin{array}{c|cccc} 1 & 0 & 0 & \cdots & 0 \\ \hline 1 & & & & \\ \vdots & & & {}^1N & \\ 1 & & & & \end{array}\right),$$

where ${}^1N$ is the fundamental matrix of the transient chain obtained by making state 1 absorbing. The interpretation of ${}^1N_{ij}$ is the mean number of times a process started in i steps into j before absorption. Here $\alpha g=\alpha_1$, and hence (9) shows that $\alpha=\alpha_1(\beta Z)$. For the first component this is an identity, but for $j\neq 1$ we obtain the identity

$$\sum_k P_{1k}\,{}^1N_{kj}=\frac{\alpha_j}{\alpha_1}. \tag{34}$$

Also for $i\neq 1$

$$(Zh)_i=1+\sum_j {}^1N_{ij}=1+M_{i1};$$

hence from (30)

$$M_{ij}=\frac{{}^1N_{jj}-{}^1N_{ij}}{\alpha_j}-M_{j1}+M_{i1}.$$

This result is correct also when i or j is 1, if we let (as usual) ${}^1N_{ij}=0$ if i or j is 1, and $M_{11}=0$. Since we could have used in place of 1 a general state k, we obtain the interesting identity

$$M_{ij}=\frac{{}^kN_{jj}-{}^kN_{ij}}{\alpha_j}-M_{jk}+M_{ik}. \tag{35}$$

Multiplying by α_j and summing on j we obtain, using $M_{ik}=\Sigma_j\,{}^kN_{ij}$,

$$\sum_j M_{ij}\alpha_j=\sum_j {}^kN_{jj}-(\alpha M)_k. \tag{36}$$

Since i does not occur on the right side, the left side is a constant (independent of i). Similarly the right side is independent of k. We have previously urged readers to try to find a probabilistic interpretation for this constant, but so far none has been found. Still another expression for this constant may be found from (30):

$$\text{const} = \sum_j M_{ij}\alpha_j = \sum_j Z_{jj} - \alpha Zh. \tag{37}$$

This result was previously known for the special case $Z = Z^*$, for which $\alpha Zh = 1$, but it holds for all our Z's. For Oz the constant is $\frac{32}{15}$, as can be computed from either Z^* or Z_β.

Our final application is to the Markov-process version of classical potential theory. Such a theory exists for both functions and measures (column and row vectors in the finite case). A charge is a function f of total integral 0 (i.e., $\alpha f = 0$) or a measure of total measure 0 (i.e., $\phi h = 0$). Potentials satisfy a certain averaging property—they are solutions of (2) or (13), respectively. We know that there always are such potentials for any charge, from Theorems 2 and 3, and that uniqueness requires an additional condition. The usual conditions imposed have been $\alpha x = 0$ and $\xi h = 0$. Hence Z^* is suitable as a potential operator for both functions and measures, and $x = Z^* f$, $\xi = \phi Z^*$.

We can generalize this theory by imposing different boundary conditions on the potentials. If we require that $\beta x = 0$ and $\xi g = 0$, then the potential operator is the Z determined by g and β, and $x = Zf$, $\xi = \phi Z$.

These generalized potentials have an amusing nonprobabilistic application. Consider n teams involved in a tournament. We wish to measure the relative strengths of the teams even though not every team has played every other team. Let s_{ij} = number of points by which team i beat team j (a negative number if j won). We wish to assign point ratings to teams, x_i. Ideally one wishes that

$$s_{ij} = x_i - x_j \tag{38}$$

for every game. But this is too much to expect; teams have "good days" and "bad days". What we will require is that for each team i, the *sum* of the differences $s_{ij} - (x_i - x_j)$ be zero. If team i has played t_i games, this means that

$$x_i = \frac{1}{t_i} \sum_j (x_j + s_{ij}), \tag{39}$$

where the sum is taken over all the opponents i has played. Let us introduce the matrix P defined to have $P_{ij}=1/t_i$ if i has played j, and 0 otherwise, and the vector f with $f_i=(1/t_i)\Sigma_j s_{ij}$. Then (39) takes on the form (1). The matrix P is nonnegative and has the constant vector h as fixed point, but is it ergodic? This will be the case if every team either has played any other team or has played teams that played teams (etc.) that have played that team. Clearly, without such a connection meaningful ratings are not possible.

The vector α is defined by $\alpha_i=t_i/t$, where $t=\Sigma t_i$, and αf is the average of the s_{ij}, which is 0, since $s_{ji}=-s_{ij}$. Thus a solution of (39) exists. We know that an extra condition may be imposed, which is not surprising, since in (38) only the differences of the ratings matter. We might decide to give one team rating 0, and rate all other teams relative to it. This would be achieved by choosing a β with 1 in that component and 0 otherwise. Or we might make the sum of the ratings equal to zero, choosing $\beta_i=1/n$. In either case we have $\beta x=0$ as our condition and compute Z_β; the ratings are then given by $x=Z_\beta f$. It is worth noting that the condition $\alpha x=0$ would be quite unnatural.

HISTORICAL NOTES

The matrix Z^* was introduced in [4] and has been widely used. Various alternatives have also been proposed. Hunter showed [2] that Z^* is a "generalized inverse" of $I-P$, in the sense that

$$(I-P)Z^*(I-P)=I-P. \tag{40}$$

He also extended the use of the fundamental matrix to Markov renewal processes.

It should be pointed out that *all* the matrices (4) are generalized inverses of $I-P$, as follows immediately from either (10) or (11).

The paper [9] compares alternative methods for calculating the vector α on a computer. The recommended method that emerges from this work is one of the Z_β matrices, with β chosen as a row of P, and α calculated as in (28). [Cf. the remark following (28).]

Campbell and Meyer [1] show that $I-P$ has a group inverse $(I-P)^{\#}$ for *any* Markov chain P, and that it can be used to calculate key quantities. For a regular chain, $(I-P)^{\#}=Z^*-h\alpha$. Thus the group inverse is very close to Z^*; indeed, on the range of $I-P$ they are the same invertible operator. Thus they are equivalent as potential operators. The similarity is not so great to the Z_β's. The range of $I-P$ is the set $\{x\,|\,\alpha x=0\}$. A Z_β, $\beta\neq\alpha$, maps this set onto the set $\{x\,|\,\beta x=0\}$. That is why these matrices are new potential operators and provide greater flexibility.

For a treatment of potential theory for Markov chains the reader is referred to [5].

REFERENCES

1 S. L. Campbell and C. D. Meyer, *Generalized Inverses of Linear Transformations*, Pitman, 1979.
2 Jeffrey J. Hunter, On the moments of Markov renewal processes, *Advances in Appl. Probability* 1:188–210 (1969).
3 John G. Kemeny, J. Laurie Snell, and Gerald L. Thompson, *Introduction to Finite Mathematics*, Prentice-Hall, 1956.
4 John G. Kemeny and J. Laurie Snell, *Finite Markov Chains*, Van Nostrand, 1960; Springer, 1976.
5 John G. Kemeny, J. Laurie Snell, and Anthony W. Knapp, *Denumerable Markov Chains*, Springer, 1976.
6 John G. Kemeny and Thomas E. Kurtz, BASIC *Programming*, 3rd ed., Wiley, 1980.
7 Carl D. Meyer, An alternative expression for the mean first passage matrix, *Linear Algebra and Appl.* 22:41–47 (1978).
8 Carl D. Meyer, The role of the group generalized inverse in the theory of finite Markov chains, *SIAM Rev.* 17:443–464 (1975).
9 C. C. Paige, George P. H. Styan, and Peter G. Wachter, Computation of the stationary distribution of a Markov chain, *J. Statist. Comp. and Simulation* 4:173–186 (1975).

Received 14 March 1980; revised 3 September 1980

Undergraduate Texts in Mathematics

continued from ii

Prenowitz/Jantosciak: Join Geometrics: A Theory of Convex Set and Linear Geometry.
1979. xxii, 534 pages. 404 illus.

Priestly: Calculus: An Historical Approach.
1979, xvii, 448 pages. 335 illus.

Protter/Morrey: A First Course in Real Analysis.
1977. xii, 507 pages. 135 illus.

Ross: Elementary Analysis: The Theory of Calculus.
1980. viii, 264 pages. 34 illus.

Sigler: Algebra.
1976. xii, 419 pages. 27 illus.

Simmonds: A Brief on Tensor Analysis.
1982. xi, 92 pages. 28 illus.

Singer/Thorpe: Lecture Notes on Elementary Topology and Geometry.
1976. viii, 232 pages. 109 illus.

Smith: Linear Algebra.
1978. vii, 280 pages. 21 illus.

Smith: Primer of Modern Analysis
1983. xiii, 442 pages. 45 illus.

Thorpe: Elementary Topics in Differential Geometry.
1979. xvii, 253 pages. 126 illus.

Troutman: Variational Calculus with Elementary Convexity.
1983. xiv, 364 pages. 73 illus.

Whyburn/Duda: Dynamic Topology.
1979. xiv, 338 pages. 20 illus.

Wilson: Much Ado About Calculus: A Modern Treatment with Applications Prepared for Use with the Computer.
1979, xvii, 788 pages. 145 illus.

9 780387 901923